Diabetes and Protein Glycosylation

Margo Panush Cohen

Diabetes and Protein Glycosylation

Measurement and Biologic Relevance

With a Foreword by Harold Rifkin

With 32 Figures

Springer-Verlag
New York Berlin Heidelberg Tokyo

Margo Panush Cohen
Professor of Medicine and Chair
Division of Endocrinology
  and Metabolism
University of Medicine
  and Dentistry of New Jersey
Newark, New Jersey 07103
U.S.A.

Library of Congress Cataloging in Publication Data
Cohen, Margo P.
  Diabetes and protein glycosylation.
  Bibliography: p.
  Includes index.
  1. Diabetes. 2. Glycoproteins. 3. Glycosylated
hemoglobin. I. Title. [DNLM: 1. Diabetes Mellitus—
metabolism. 2. Glycoproteins—biosynthesis.
WK 810 C678d]
RC660.C473 1986   616.4'62   86-3949

Typeset by Ampersand Publisher Services Inc., Rutland, Vermont.
Printed and bound by R.R. Donnelley & Sons, Harrisonburg, Virginia.
Printed in the United States of America.

9 8 7 6 5 4 3 2 1

ISBN-13: 978-1-4612-9366-8   e-ISBN-13: 978-1-4612-4938-2
DOI: 10/1007-978-1-4612-4938-2

To Louis and Tillie Panush
and to Perry, Michael, Daniel,
and Jonathan Cohen

Foreword

In the years since the initial discovery that blood from diabetic patients contains increased amounts of a posttranslationally glucosylated form of hemoglobin (hemoglobin A_{1c}), an impressive number of studies have clarified and expanded the use of glycohemoglobin levels to assess disease status. Many other structural proteins have been shown to undergo similar changes, including proteins from tissues most commonly affected in diabetes (e.g., lens, aorta, peripheral nerve, basement membrane). Thus, the nonenzymatic glycosylation of hemoglobin emerges as an invaluable model for the pathogenesis of certain chronic diabetes complications.

In addition to reviewing a wealth of investigative possibilities in the area of these chronic complications—including eye, kidney, nerve, and vascular disease—Dr. Cohen indicates how enhanced nonenzymatic glycosylation in uncontrolled diabetes underscores the pressing need for maintenance of long-term euglycemia.

Dr. Cohen is an endocrinologist and diabetes specialist whose research activities have largely focused on the chemistry and metabolism of the basement membrane in diabetes. This superb monograph on nonenzymatic glycosylation clearly shows the major trends of her past and present research and clinical activities.

This book is beautifully written and a pleasure to read. It provides great insight into the mechanisms of the pathogenesis of the complications of diabetes and should be of immense value not only to basic and clinical investigators, but also to internists, diabetologists, and endocrinologists in clinical practice.

HAROLD RIFKIN

Clinical Professor of Medicine
Albert Einstein College of Medicine

Professor of Clinical Medicine
New York University School of Medicine

Principal Consultant
Diabetes Research and Training Center
Albert Einstein College of Medicine
Montefiore Medical Center
New York, New York

Preface

Nonenzymatic glycosylation of proteins, long known to food chemists as the "browning" reaction, has become the subject of a surge of interest that has made its biologic relevance—particularly in diabetes—increasingly clear in recent years. Hyperglycemia promotes the enhanced nonenzymatic glycosylation of both circulating and tissue proteins, thereby not only allowing the assessment of diabetes control through determination of glycohemoglobin and glycoalbumin levels, but also providing insight into the pathogenetic mechanisms associated with the chronic complications of diabetes, as well as with the aging process.

Diabetes and Protein Glycosylation: Measurement and Biologic Relevance was written with the conviction that the crucial role of glucose in the pathogenesis of diabetic complications can no longer be ignored. A comprehensive discussion of the chemistry, methods for measurement, clinical use and abuse, and pathophysiologic consequences of nonenzymatic protein glycosylation should be of value to all physicians concerned with the diagnosis and management of diabetes and to clinical and basic science investigators interested in the biochemical basis of the disease's sequelae.

New York, New York
February 1986

M.P. COHEN

Contents

CHAPTER 1

Introduction

Recent years have witnessed a surge of interest in nonenzymatic glycosylation, which is the attachment of free sugar to certain amino acid residues of proteins. It has become increasingly clear that this reaction, long known to food chemists as the "browning" reaction that occurs in baked or stored foodstuffs,[1-3] has considerable biomedical relevance, particularly in diabetes. Hyperglycemia promotes increased nonenzymatic glycosylation not only of circulating proteins such as hemoglobin and albumin, thus allowing measurement of glycohemoglobin or glycoalbumin to be used for assessing diabetic control, but also of tissue proteins, thereby providing insight into pathogenetic mechanisms contributory to chronic complications of diabetes.

The modern era of interest in the potential clinical application of measurement of modified hemoglobins began with the observation that blood from patients with diabetes contained increased amounts of an unusual hemoglobin that migrated faster than hemoglobin A on electrophoresis or with carboxymethylcellulose chromatography.[4-6] The identification of this species as hemoglobin A_{1c}, and the recognition that it represented a postribosomal modification formed by the attachment of glucose to the amino-terminal of the β chain, attracted the attention of clinicians and investigators concerned with the management of diabetes. Interest was further spurred by the demonstration that nonenzymatic glycosylation of hemoglobin in vivo

1

occurred slowly and cumulatively, that the proportion of total hemoglobin that is glycosylated is increased in diabetic patients with attendant hyperglycemia, and that measurement of the percentage of glycosylated hemoglobin in an individual patient could reflect ambient glucose concentrations over an integrated period of time.[7,8] The years since have witnessed an explosion of reports concerning the use of glycohemoglobin levels as a tool for assessing diabetic status, the development of new and improved methodology for rapid and reproducible determinations, a further understanding of the nature and sites of hemoglobin glycosylation, and the advent of widespread use of glycohemoglobin measurements as a parameter for estimating glycemic control in diabetic patients.

The hypothesis that nonenzymatic glycosylation of hemoglobin might be a model reaction relevant to the pathogenesis of certain complications of chronic diabetes emerged with the demonstration that a variety of other proteins are also subject to glycosylation under physiologic conditions, and soon attracted the interest of investigators concerned with the biochemical basis of diabetic sequelae.[9,10] The list of proteins known to undergo increased nonenzymatic glycosylation in patients or animals with diabetes mellitus is growing, and the reaction has been shown to proceed with virtually every protein that has been examined. Many of the proteins studied derive from specific tissues typically involved with diabetic complications, such as basement membrane, aorta, lens, and peripheral nerve. With the mounting evidence that nonenzymatic glycosylation can alter the structure and/ or function of involved proteins (Table 1-1), the notion that the reaction participates in the development of complications appears more convincing and has attracted more adherents. This has occurred during a period in which it has become increasingly clear that hyperglycemia per se exerts a deleterious influence on several metabolic processes implicated in the pathogenesis of diabetic angiopathy, including basement membrane synthesis and turnover, the polyol pathway, and *myo*-inositol metabolism. Thus, the guilt of glucose can no longer be denied, and has prompted an awareness of the importance of long-term maintenance of euglycemia in diabetic patients in an attempt to prevent the development or arrest the progression of chronic complications. With respect to the potential contribution of nonenzymatic glycosylation in this scheme, the need for long-term control is further underscored by the recognition that proteins in some of the affected tissues have long biologic half-lives, and hence represent situations in which glycosylation could be relatively permanent since the population of proteins is not quickly replaced. For example, since the major lens crystallins remain throughout life, the amount of glycosylated crystallins would be

TABLE 1-1 Sample Consequences of Excess Nonenzymatic Glycosylation of Proteins

Protein	Consequence
Hemoglobin	Increased oxygen affinity
Serum albumin	Increased transendothelial transport
Low-density lipoproteins	Decreased receptor-mediated uptake and degradation
Lens crystallins	Aggregation
Collagen	Decreased solubility, abnormal cross-links; immunogenicity
Fibronectin	Decreased ligand binding
Myelin	Macrophage recognition

expected to gradually and continuously increase in the presence of hyperglycemia, until equilibrium is reached.

In short, the recognition that free sugar condenses nonenzymatically with proteins and that this reaction is increased when hyperglycemia attends the diabetic state has offered improved means for assessing diabetic control and has opened new avenues of investigation concerning the pathogenesis of diabetic complications. The chapters that follow review the chemistry, methods for measurement, clinical relevance, physiologic consequences, and pathophysiologic significance of the nonenzymatic glycosylation of proteins. Because of the importance that measurement of glycohemoglobin levels has assumed in following diabetic control, and because a variety of methods are available, procedures and caveats concerning these laboratory determinations are presented in detail. An understanding of what the various methods measure, and of the factors that influence them, is essential for proper interpretation of the results. A comprehensive discussion of studies examining the impact of nonenzymatic glycosylation on structural and functional properties of proteins is also presented, with the belief that cognizance of this information will reinforce the conviction that every effort should be made to achieve and maintain normalization of blood glucose levels in diabetic patients.

References

1. Mohammad A, Olcott HS, Fraenkel-Conrat H: The "browning" reaction of proteins with glucose. *Arch Biochem* 1979;24:157–163.
2. Reynolds TM: Chemistry of nonenzymatic browning. *Adv Food Res* 1963; 14:167–277.

3. Adhikari HR, Tappel AL: Fluorescent products in a glucose-glycine browning reaction. *J Food Sci* 1973;38:486–488.
4. Rabhar S: An abnormal hemoglobin in red cells of diabetics. *Clin Chim Acta* 1968;22:296–298.
5. Rabhar S, Blumenfeld O, Ranney HM: Studies of the unusual hemoglobin in patients with diabetes mellitus. *Biochem Biophys Res Commun* 1969; 36:833–843.
6. Trivelli LA, Ranney HM, Lai HT: Hemoglobin components in patients with diabetes mellitus *N Engl J Med* 1971;284:353–357.
7. Bunn HF, Haney DN, Kamin S, et al: The biosynthesis of human hemoglobin A_{1c}: Slow glycosylation of hemoglobin in vivo. *J Clin Invest* 1976;57:1652–1659.
8. Koenig RJ, Peterson CM, Jones RL, et al: The correlation of glucose regulation and hemoglobin A_{1c} in diabetes mellitus. *N Engl J Med* 1976;295:417–420.
9. Koenig RJ, Cerami A: Synthesis of hemoglobin A_{1c} in normal and diabetic mice: Potential model of basement membrane thickening. *Proc Natl Acad Sci USA* 1975;72:3687–3691.
10. Koenig RJ, Cerami A: Hemoglobin A_{1c} and diabetes mellitus. *Annu Rev Med* 1980;31:290–234.

CHAPTER 2

Chemistry

Condensation of Glucose With Proteins

Nonenzymatic glycosylation is a condensation reaction between carbohydrate and free amino groups at the NH_2-terminus or ε-amino groups of lysine residues of proteins. The reaction is initiated with attachment of the aldehyde function of acyclic glucose to a protein amino group via nucleophilic addition, forming an aldimine, also known as a Schiff base (Figure 2-1). This intermediate product subsequently undergoes an Amadori rearrangement to form a 1-amino-1-deoxyfructose derivative in stable ketoamine linkage, which in turn can cyclize to a ring structure.[1-6] This bimolecular condensation of free saccharide with protein constitutes a mechanism by which proteins are subjected to postribosomal modification without the influence of enzymatic activities.

Outside of the food science industry, the chemistry of the non-enzymatic glycosylation reaction as it concerns biologically relevant proteins is best described for glucose and hemoglobin. However, it is clear that a host of other proteins are subject to nonenzymatic glycosylation both in vivo and in vitro, and that sugars other than glucose can attach to proteins (see the following section). Nevertheless, glucose and hemoglobin remain the prototypical reactants with which

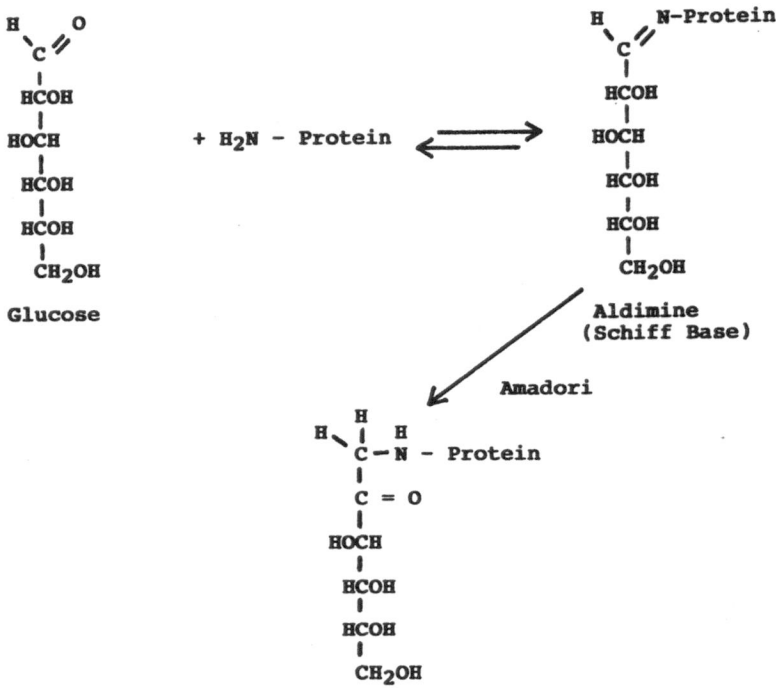

FIGURE 2-1   Reaction pathway for nonenzymatic glycosylation of proteins.

many of the studies concerning the chemistry and biosynthesis of nonenzymatically glycosylated proteins have been conducted.

The major product of the reaction between glucose and hemoglobin is hemoglobin A_{1c} (Hb A_{1c}), which is identical to Hb A except that glucose is linked to the amino-terminal valine residue of the β chain.[5,7-9] The labile Schiff base that is intermediate to the formation of Hb A_{1c} has been called "pre-A_{1c}" and increases rapidly in amount within a few hours after incubation of Hb A with glucose, until equilibrium is reached.[10] In contrast, formation of Hb A_{1c}, the stable ketoamine, is negligible during this time. The rate of Amadori rearrangement has been calculated to be about 1/60th of the rate of dissociation of the aldimine back to glucose and hemoglobin, indicating that the rearrangement step is rate limiting for the formation of Hb A_{1c}. In vivo, Hb A_{1c} is formed slowly and continuously throughout the life of the red cell.[11-13] After infusion of ^{59}Fe-labeled transferrin in normal individuals or animals, the specific activity of Hb A_{1c} (as well as of Hb A_{1a} and Hb A_{1b}) progressively increased over the ensuing 80 to 100 days and exceeded that of Hb A after about 60 days (Figure 2-2). This is compatible with a slow and irreversible process.

Of parenthetical interest are two recent papers that propose another manner, independent of the traditional concept of nonenzymatic glycoslation, by which Hb A_{1c} can be formed.[14,15] These reports describe the ability of an erythrocyte membrane glucose-containing factor, believed to be a glycoconjugate and soluble in neutral buffer and chloroform, to transfer glucose to hemoglobin in vitro. Extracts from young but not old red blood cell membranes could accomplish transfer of glucose to Hb A. The investigators suggested that Hb A_{1c} can be formed by two complementary mechanisms: one that involves glucose transfer from a membrane glycoconjugate and is operative during the first days of red blood cell intravascular life, and another that involves condensation of free glucose and slowly continues during the full life span of the erythrocyte.

Three other minor components of hemoglobin, in addition to Hb A_{1c}, are modified at the NH_2-terminus of the β chain.[16,17] These have

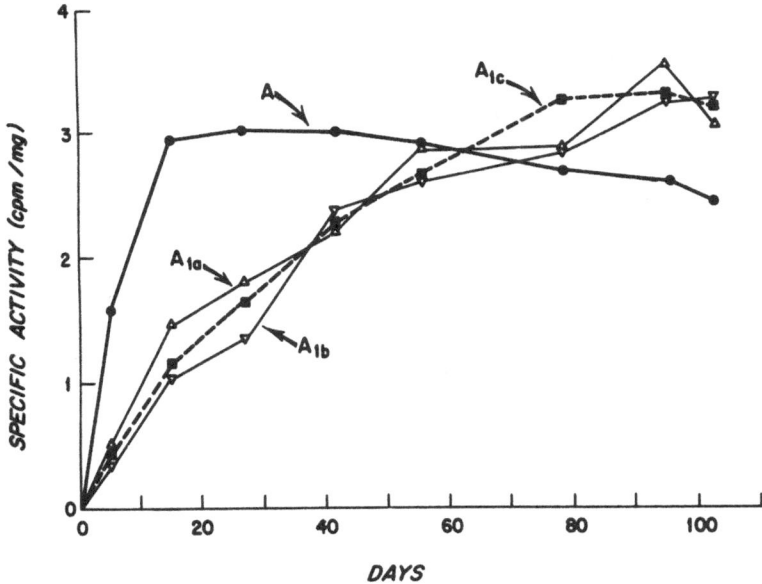

FIGURE 2-2   In vivo radiolabeling of Hb A and minor hemoglobins A_{1a-c} after infusion of [^{59}Fe] transferrin. Reproduced from *The Journal of Clinical Investigation*, 1976,57:1652–1659, by copyright permission of The American Society for Clinical Investigation.

been designated Hb A_{1a_1}, Hb A_{1a_2}, and Hb A_{1b}. Like Hb A_{1c}, the Hb A_{1a} components contain carbohydrate in ketoamine linkage, and they each comprise about 0.2% of total hemoglobin (Table 2-1). Hemoglobins A_{1a_1} and A_{1a_2} contain phosphate as well as carbohydrate; there are two phosphates per $\alpha\beta$ dimer in Hb A_{1a_1} and one per dimer in Hb A_{1a_2}. This would be consistent with modification, presumably by condensation reaction, by a bisphosphate sugar such as fructose-1,6-bisphosphate to form Hb A_{1a_1}, and by a monophosphate sugar such as glucose-6-phosphate to form Hb A_{1a_2}.[16,18,19] Structure assignments for Hb A_{1b} are not complete, but it may be a deamidization product of Hb A.[20] Alternatively, it may represent a further posttranslational modification of Hb A_{1c}, as suggested by results of in vivo biosynthetic studies in which ^{59}Fe-labeled transferrin was infused into rhesus monkeys.[13] The interpretation that Hb A_{1a_1} and Hb A_{1a_2} represent adducts of sugar phosphates is consistent with the observation that fructose-6-phosphate, fructose-1,6-diphosphate, and glucose-6-phosphate react readily with hemoglobin in vitro.[18,19,21]

In addition to the NH_2-terminus of the β chain, glucose adducts can form with the amino-terminal of the α chain as well as with other free amino groups in the hemoglobin molecule. Various lysine residues in the α and β chains become glycosylated on exposure to glucose not only in vitro but also in vivo.[22,23] It has been estimated that about 8% to 10% of Hb A_0 is glycosylated at the amino-termini of the α chains or at lysine amino groups in the α and β chains. Further, it is now abundantly evident that glucose can condense with ε-amino groups of lysine (or hydroxylysine) residues along the polypeptide chains of a wide variety of proteins. Like Hb A_{1c}, the amount of glycosylated Hb A in red cells from diabetic patients is increased[24]; unlike Hb A_{1c}, however, modification by glucose at these other sites does not result in changes in electrophoretic or ion-exchange chromatographic properties.

TABLE 2-1   Minor Components of Hemoglobin A

Hemoglobin	Modification	Abundance (%)
A ($\alpha_2\beta_2$)	...	95*
A_{1a_1}	$\alpha_2(\beta\text{-}N\text{-fructose-1,6-diphosphate})_2$(?)	0.2
A_{1a_2}	$\alpha_2(\beta\text{-}N\text{-glucose-6-phosphate})_2$	0.2
A_{1b}	$\alpha_2(\beta\text{-}N\text{-carbohydrate})_2$(?)	0.5
A_{1c}	$\alpha_2(\beta\text{-}N\text{-glucose})_2$	4

*An estimated 8% to 10% is glycosylated at sites other than N termini.[22,23]

TABLE 2-2   Relative Reactivity of Selected
Monosaccharides With Hemoglobin*

Sugar	Relative Reactivity
Glucose	1
Galactose	4.7
Mannose	5.3
Ribose	6.7

*Data from Stevens et al.[19] and Bunn and Higgins.[26]

Other Monosaccharides

The nonenzymatic glycosylation reaction is not limited to glucose, and
a variety of sugars and sugar phosphates including ribose, galactose,
and glucose-6-phosphate can covalently attach to proteins.[18,19,21,25-28]
Glycosylation is therefore a generic term, whereas glucosylation refers
specifically to the condensation of glucose with proteins. Ribose is one
of the most effective, whereas glucose is the least effective in
glycosylation (Table 2-2), and the rates of reaction of different
monosaccharides with protein (using hemoglobin as the prototypical
model) can vary over a 300-fold range. Aldohexoses as well as
ketohexoses, such as fructose and xylulose, can condense with protein,
although in general the rate of interaction of the latter group is less
than that of the former. This appears to be due to the fact that aldehyde
carbonyl groups are relatively more electrophilic than ketone carbonyl
groups, and the reaction proceeds by a nucleophilic attack of an
unprotonated amino group on the carbonyl group of the sugar. The
relative reactivity of individual monosaccharides is strongly influenced by the extent to which the compound exists in the open
(acyclic) form, which in turn depends on the equilibrium between the
open and ring configurations.

   In the context of this discussion, it is interesting to note that the level
of glycosylated hemoglobin has been found to be increased in patients
with glucose-6-phosphate dehydrogenase (G6PD) deficiency.[29] The
mean Hb A_{1c} in 17 GGPD-deficient Libyan subjects was significantly
greater than that in 34 matched controls, and 6 of the 17 G6PD-
deficient men had levels that were 2 or more standard deviations above
the normal mean. The explanation offered for this finding is that
erythrocyte proteins, including hemoglobin, are nonenzymatically
glycosylated by glucose-6-phosphate, which accumulates as a result of
the enzymatic defect.[30] This interpretation seems valid in light of the
fact that plasma glucose concentrations are not elevated in G6PD-
deficient individuals.

Although nonenzymatic glycosylation can involve different sugars, glucose is the major carbohydrate nutrient in humans. In view of the mounting evidence that the reaction exerts deleterious effects on structure-function relationships of involved proteins, it is of heuristic interest to consider factors that protect humans from excess non-enzymatic glycosylation and its potentially injurious consequences. According to Bunn and Higgins,[26] this relates to the stability of the ring structure of glucose, which, under normal circumstances and at physiologic concentrations of glucose, limits the forward reaction that attaches glucose to proteins. The rate of reaction of galactose, on the other hand, is several times that of glucose as a result of the greater proportion of galactose that exists in the open form. Hence, the extent of nonenzymatic glycosylation with galactosemia is considerably greater than that occurring with a comparable degree of hyper-glycemia. Thus, Bunn and Higgins proposed that the emergence of glucose rather than other monosaccharides as the most important metabolic fuel coincided with the stability of its ring structure, which limits undesirable side reactions attributable to covalent modification of proteins via nonenzymatic glycosylation.

Modifying Factors

Nonenzymatic glucosylation takes place under physiologic conditions in normal individuals. The reaction follows second-order kinetics, and the amount of glucosylated product is proportional to the concentration of reactants. Since the concentration of proteins remains fairly stable, a major determinant of the level of glucosylation is the glucose concentration, which will, according to the law of mass action, cause a proportionate increase in the amount of Amadori product formed. This relationship has been demonstrated repeatedly in vitro and pertains even with very high glucose concentrations.[31] The second main determinant is the time of exposure of the protein to increased glucose concentration, since glucoadducts will continue to form as a function of time until equilibrium is reached. In vivo, these two influences (glucose concentration and time) translate to degree and duration of hyperglycemia (Table 2-3).

Several other factors influence both the absolute amount and the percentage of total of a particular protein that becomes glucosylated. Some, such as pH and temperature, are primarily relevant in vitro. Others are of substantial significance in vivo, particularly from the perspective of the potential links among chronic hyperglycemia, nonenzymatic glycosylation, and complications of diabetes. Long-lived proteins may be especially vulnerable by virtue of the temporal

TABLE 2-3   Factors Influencing Extent of Glucosylation in Vivo

Degree and duration of hyperglycemia
Half-life of the protein in the circulation or in tissue
Permeability of tissue to free glucose
Number of free amino groups
Accessibility and pK of the amino groups within the structure of the protein

factor, which heightens the possibility that they will be modified. Because the reaction products are cumulative, the level of glucosylation of a protein with a slow turnover, such as collagen, may increase with age despite normal blood glucose concentration; in the presence of diabetes, the amount of glucose nonenzymatically bound to tendon collagen is further increased.[32] Similarly, there is an age-related increase in the amount of ketoamine-linked glucose bound to insoluble collagen; the extent of glucosylation of insoluble skin collagen in juvenile diabetic subjects is greater than that in comparably aged control subjects.[33] On the other hand, reduction in the half-life of a protein or in its residence time in the circulation can decrease the extent of glucosylation even without a correspondent change in blood glucose levels. A good illustration of this is the decreased content of Hb A_{1c} in erythrocytes of patients with hemolytic anemia and a shortened red cell life span, and in erythrocytes of normal subjects who have sustained an acute blood loss.[11,34] Conversely, old erythrocytes contain more Hb A_{1c} than do young ones.[15,35]

The permeability and availability of glucose in different tissues will clearly influence the extent of nonenzymatic glucosylation of proteins in these tissues. An interesting example is pig hemoglobin; since glucose is not transported into pig erythrocytes, the level of glucosylated hemoglobin in red cells of this species is extremely low.[36] Human erythrocytes, in contrast, are freely permeable to glucose, and hence their glucose concentration would be expected to mirror that in the vascular compartment. In fact, among the various species examined, the highest glycosylated hemoglobin levels were found in humans.[37] On the other hand, the rate of glycosylation of proteins that are freely exposed to glucose in the blood, such as albumin, is similar in a variety of species.[37] It is of more than passing interest to note that several characteristic complications of diabetes occur in tissues that do not require insulin for glucose transport and can dispose of glucose via insulin-independent pathways. This observation increases the attractiveness of the concept, discussed in later sections, that excess nonenzymatic glycosylation contributes to the pathogenesis of diabetic complications in these tissues.

Any free amino group is a potential site for nonenzymatic glycosylation, and one would expect a priori that there would be increased glycosylation as the amount of protein, and hence the absolute number of free amino groups, increases. However, the reactivity of any individual free amino group in a given protein is not uniform and depends on its accessibility, its pK within the structure of the protein, and microenvironmental factors such as the availability of nearby catalytic carboxyl groups for acceleration of Amadori rearrangement. This may explain the apparent preferential glycosylation of various residues in a given protein, the fact that some proteins can become more extensively glycosylated than others, and that there are differences between the sites and order of prevalence of glycosylation occurring in vivo versus in vitro. For example, although the NH_2-terminal amino group of the β chain of hemoglobin is the most reactive group both in vivo and in vitro, several other sites on the β chain as well as on the α chain can be modified.[22] Incubation of purified Hb A_0 with [^{14}C]glucose followed by ion-exchange chromatography and two-dimensional peptide mapping allowed identification of the major sites of glucosylation occurring in vitro as β-Val-1, α-Lys-16, β-Lys-66, β-Lys-17, α-Val-1, αLys-7, and β-Lys-120. In contrast, the major sites of glucosylation occurring in vivo were identified, in order of prevalence, as β-Val-1, β-Lys-66, α-Lys-61, β-Lys-17, and α-Val-1.[23] Hemoglobin from diabetic subjects contains, in addition to Hb A_{1c}, glucoadducts formed with lysine residues of the β chain and the α chain, and with the NH_2-terminal valine of the α chain.[38] When human erythrocytes are incubated under physiologic conditions for several days with high glucose concentrations (1,000 mg/dL), the rate of formation of Hb A_{1c} is increased by increasing the concentration of 2,3-diphosphoglycerate.[39] Additionally, deoxyhemoglobin is glucosylated twice as rapidly as oxyhemoglobin, and the effect of diphosphoglycerate on glucosylation is greatest when hemoglobin is in the deoxygenated state. These findings suggest that the reactivity of the NH_2-terminal valine of the β chain is influenced by the conformation of the protein and/or by local shifts in charge.

Maillard Browning

The ketoamine adduct formed from the reaction between glucose and protein amino groups can undergo a series of dehydration, rearrangement, and cleavage reactions collectively known as the Maillard or browning reaction.[40,41] The products of such reactions are highly cross-linked, insoluble, pigmented, and fluorescent carbohydrate-protein polymers called melanoidins. In the past, their formation has been most extensively studied in vitro and in foodstuffs, but recent evidence indicates that analogous products are formed in vivo, particularly in long-lived proteins and especially when nonenzymatic glycosylation is increased. Indeed, advanced glycosylation end products have been invoked in the pathogenesis of diabetic complications (see Chapter 5) and in the aging process, and consequences of nonenzymatic browning may be common to both aging and diabetes, which some consider as a disease causing accelerated aging.[42]

Conditions that promote the development of nonenzymatic browning products in foods include heat, storage conditions, and atmospheric oxygen, and the products themselves give rise to changes in the aesthetic, functional, and nutritional properties of stored foods.[43,44] The presence of brownish pigments makes food unattractive, whereas effects on nutritional properties relate to a decrease in the biologic availability of essential amino acids, primarily lysine but also tryptophan, arginine, and histidine.[44,45] Effects on physiologic properties include decreased digestibility. It is now appreciated that pigmented or fluorescent compounds form in vivo, and that the susceptibility to enzymatic digestion of certain "browned" proteins in tissues of diabetic subjects is decreased. Thus, insights gained from food science studies of the Maillard reaction have been useful for predicting potential consequences of the reaction if it occurs in living organisms.

Maillard browning involves three phases, the first two of which are sequential whereas the third probably consists of parallel reactions.[1,43,44] Formation of the colorless deoxyketose derivative (Amadori compound) via Schiff-base intermediate, discussed in the preceding sections, is the first step and is sometimes referred to as the "early Maillard" reaction. The next phase involves removal of amino groups via 1,2- or 2,3-enolization reactions followed by dehydration, cyclization, and fission reactions that generate secondary products such as hydroxymethylfurfural, reductones, aldehydes, and dicarbonyls. In the third phase, polymerization reactions occur, yielding nitrogen-free brown pigments if the complexing second-phase products derive from the sugar moiety of the Amadori compound, and nitrogen-containing polymers when secondary products react with amino groups to form aldimines and ketimines (Figure 2-3).

Evidence indicating that the early Maillard reaction, which is nonenzymatic glycosylation, occurs in vivo is overwhelming. That advanced Maillard reactions also can occur in vivo has only recently been appreciated. Although the detailed chemistry of these reactions and how they proceed in vivo has not been elucidated, it is clear that sequences analogous to those summarized above are probable. The biologic significance of advanced glycosylation end-product form-

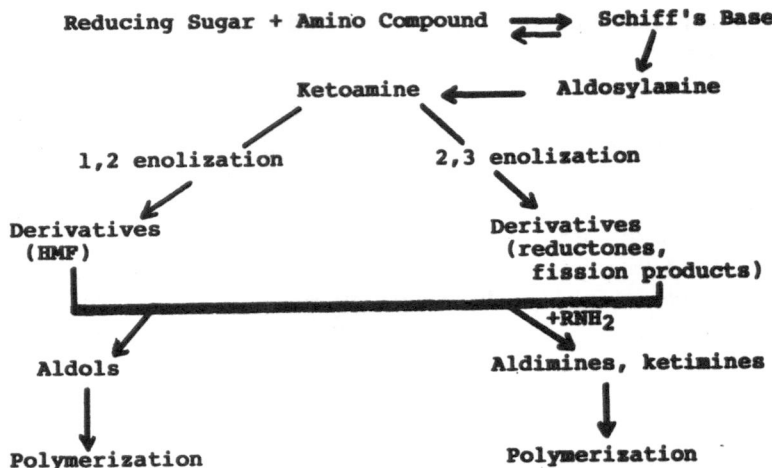

FIGURE 2-3   Simplified scheme of the Maillard reaction. HMF, 5-hydroxy-methylfurfuraldehyde. Adapted from Finot.[44]

ation and its relevance to complications of diabetes are discussed in the final chapter.

References

1. Hodge JE: Chemistry of browning reactions in model systems. *J Agric Food Chem* 1953;1:928–943.
2. Koenig RJ, Blobstein SH, and Cerami A: Structure of carbohydrate and hemoglobin A_{1c}. *J Biol Chem* 1977;252:2992–2997.
3. Fisher RW, Winterhalter KH: The carbohydrate moiety in hemoglobin A_{1c} is present in the ring form. *FEBS Lett* 1981;135:145–147.
4. Bunn HF, Gabbay KH, Gallop PM: The glycosylation of hemoglobin: Relevance to diabetes mellitus. *Science* 1978;200:21–27.
5. Bunn HF, Haney DN, Gabbay KH, et al: Further identification of the nature and linkage of the carbohydrate in hemoglobin A_{1c}. *Biochem Biophys Res Commun* 1975;67:103–109.
6. Neglia CI, Cohen HJ, Garber AR, et al: C NMR investigation of nonenzymatic glucosylation of protein: Model studies using RNase A. *J Biol Chem* 1983;258:14279–14283.
7. Bookchin RM, Gallop PM: Structure of hemoglobin A_{1c}: Nature of the N-terminal β-chain blocking group. *Biochem Biophys Res Commun* 1968;32:86–93.
8. Flückiger R, Winterhalter KG: In vitro synthesis of hemoglobin A_{1c}. *FEBS Lett* 1976;71:356–360.

9. Holmquist WR, Schroeder WA: A new N-terminal blocking group involving a Schiff base in hemoglobin A $_{1c}$. *Biochemistry* 1966;5:2489–2503.

10. Higgins PJ, Bunn HF: Kinetic analysis of the nonenzymatic glycosylation of hemoglobin. *J Biol Chem* 1981;256:5204–5208.

11. Bunn HF, Haney DN, Kamin S, et al: The biosynthesis of human hemoglobin A_{1c}: Slow glycosylation of hemoglobin in vivo. *J Clin Invest* 1976;57:1652–1659.

12. Koenig RJ, Cerami A: Synthesis of hemoglobin A_{1c} in normal and diabetic mice: Potential model of basement membrane thickening. *Proc Natl Acad Sci USA* 1975;72:3687–3691.

13. Solway J, McDonald M, Bunn HF, et al: Biosynthesis of glycosylated hemoglobin in the monkey. *J Lab Clin Med* 1979;93:962–973.

14. Gillery P, Maquart FX, Gattegno L, et al: A glucose-containing fraction extracted from young erythrocyte membrane is capable of transferring glucose to hemoglobin in vitro. *Diabetes* 1982;31:371–374.

15. Gillery P, Maquart FX, Corcy JM, et al: A glucose transfer from membrane glycoconjugates to hemoglobin in isolated young red blood cells: another biosynthetic way for glycosylated haemoglobins. *Eur J Clin Invest* 1984;14:317–322.

16. McDonald MJ, Shapiro R, Bleichman M, et al: Glycosylated minor components of human adult hemoglobin. *J Biol Chem* 1978;253:2327–2332.

17. Garrick LM, McDonald MJ, Shapiro R, et al: Structural analysis of the minor human hemoglobin components: Hb A_{1a_1}, Hb A_{1a_2} and Hb A_{1b}. *Eur J Biochem* 1980;106:353–359.

18. Haney DN, Bunn HF: Glycosylation of hemoglobin in vitro: affinity labelling by glucose-6-phosphate. *Proc Natl Acad Sci USA* 1976;73:3534–3538.

19. Stevens VJ, Vlassara H, Abati A, et al: Nonenzymatic glycosylation of hemoglobin. *J Biol Chem* 1977;252:2998–3002.

20. Krishnamoorthy R, Gacon G, Labie D: Isolation and characterization of hemoglobin A_{1b}. *FEBS Lett* 1977;77:99–101.

21. Dolhofer R, Wieland OH: In vitro glycosylation of hemoglobin by different sugars and sugar phosphates. *FEBS Lett* 1978;85:86–90.

22. Bunn HF, Shapiro R, McManus M, et al: Structural heterogeneity of human hemoglobin A due to nonenzymatic glycosylation. *J Biol Chem* 1979;254:3892–3898.

23. Shapiro R, McManus MJ, Zalut C, et al: Sites of nonenzymatic glycosylation of human hemoglobin A. *J Biol Chem* 1980;255:3120–3127.

24. Gabbay KH, Sosenko JM, Banuchi GA, et al: Glycosylated hemoglobins: Increased glycosylation of hemoglobin A in diabetic patients. *Diabetes* 1979;28:337–340.

25. Schwimmer S, Olcott HS: Reaction between glycine and the hexose phosphates. 02J Am Chem Soc 1953;75:4855–4856.

26. Bunn HF, Higgins PJ: Reaction of monosaccharides with proteins: Possible evolutionary significance. *Science* 1981;213:222–224.

27. Kohn RR, Cerami A, Monnier VM: Collagen aging in vitro by non-enzymatic glycosylation and browning. *Diabetes* 1984;33:57–59.

28. Urbanowski JC, Cohenford MA, Dain JA: Nonenzymatic glycosylation of human serum albumin: In vitro preparation. *J Biol Chem* 1982;257:111–115.

29. Donde UM, Baxi AJ, El Tawil H, et al: Glycosylated haemeoglobin in glucose-6-phosphate dehydrogenase deficiency, letter to the editor. *Lancet* 1985;2:47.

30. Bernstein RE: Glycosylated haemoglobin in glucose-6-phosphate dehydrogenase deficiency, letter to the editor. *Lancet* 1985;2:332–333.

31. Nahum, HD, Lonchampt M, Duhault J: Nonenzymatic plasma protein glycosylation: Lack of saturation at high glucose concentration. *IRCS Med Sci* 1982;10:436.

32. Schnider SL, Kohn RR: Glucosylation of human collagen in aging and diabetes mellitus. J Clin Invest 1980;66:1179–1181.

33. Schnider SL, Kohn RR: Effects of age and diabetes mellitus on the solubility and nonenzymatic glucosylation of human skin collagen. *J Clin Invest* 1981;67:1630–1635.

34. Starkman HS, Wacks M, Soeldner JS, et al: Effect of acute blood loss on glycosylated hemoglobin determinations in normal subjects. *Diabetes Care* 1983;6:291–294.

35. Fitzgibbons JF, Koler RD, Jones RT: Red cell age-related changes of hemoglobin A_{1a+b} and A_{1c} in normal and diabetic subjects. *J Clin Invest* 1976;58:820–824.

36. Higgins PJ, Garlick RL, Bunn HF: Glycosylated hemoglobin in human animal red cells: Role of glucose permeability. *Diabetes* 1982 31:743–748.

37. Rendell M, Stephen PM, Paulsen R, et al: An interspecies comparison of normal levels of glycosylated hemoglobin and glycosylated albumin. *Comp Biochem Physiol* 1985;81B:819–822.

38. Garlick RL, Mazer JS, Higgins PJ, et al: Characterization of glycosylated hemoglobin. Relevance to monitoring of diabetic control and analysis of other proteins *J Clin Invest* 1983;71:1062–1072.

39. Smith RJ, Koenig RJ, Binnerts A, et al: Regulation of hemoglobin A_{1c} formation in human erythrocytes in vitro. Effects of physiologic factors other than glucose. *J Clin Invest* 1982;69:1164–1168.

40. Maillard LC: Action des acides aminés sur les sucres; formation des melanoides par voie methodique. *CR Acad Sci* 1912;154:66–68.

41. Maillard LC: Synthése des matiéres humiques par action des acides aminés sur les sucres reducteurs. *Ann Chim* 1916;6:258–317.

42. Bunn HF: Non-enzymatic glycosylation of protein: A form of molecular aging. *Schweiz Med Wochenschr* 1981;111:1503–1507.

43. Saltmarch M, Labuzza TP: Nonenzymatic browning via the Maillard reaction in foods. *Diabetes* 1982;31(suppl 3):29–36.

44. Finot PA: Nonenzymatic browning products: Physiologic effects and metabolic transit in relation to chemical structure. A review. *Diabetes* 1982;31(suppl 3):22–28.

45. Finot PA, Bujard E, Mottu F, et al: Availability of the true Schiff bases of lysine: Chemical evaluation of the Schiff base between lysine and lactose in milk. *Adv Exp Med Biol* 1977;86:343–365.

CHAPTER 3

Measurement

Circulating Proteins

Hemoglobin

Early techniques for the separation of minor hemoglobin components utilized starch block electrophoresis and carboxymethylcellulose chromatography.[1,2] These techniques not only provided means for separation on minor components, but were also instrumental in their identification and in the development of a system of nomenclature. Ion-exchange chromatography remains one of the most commonly employed methods for separation of minor hemoglobin fractions from hemoglobin A, but the original tedious techniques have been replaced in most clinical chemistry laboratories by commercially available minicolumn systems that are marketed as kits complete with pre-packed columns, prepared buffers, and lyophilized standards.[3-6] Affinity chromatography on phenylboronate, also commercially marketed as minicolumns in prepackaged kits, have become popular for measurement of glycosylated hemoglobin.[7-14] Another simple and widely employed technique entails measurement of the colored product after reaction with thiobarbituric acid (TBA)[15-21]; fluorometric detection of formaldehyde released after periodate oxidation of hemoglobin-linked carbohydrate groups is an alternative method.[22]

Several sophisticated methods, perhaps more suitable for the investigative laboratory than for routine clinical chemistry, have also been developed. These include isoelectric focusing, high-pressure liquid chromatography (HPLC), agar gel electrophoresis, and radioimmunoassay[23-27] However, as exemplified by the recent commercial appearance of instrumentation for an automated HPLC system,[28] purported to allow faster and better resolution, the use of some of these methods in the clinical laboratory may be increasing.

Reaction With Thiobarbituric Acid

This colorimetric procedure is based on the formation of a chromogen upon reaction with TBA of the 5-hydroxymethylfurfuraldehyde (HMF) that is generated after oxalic acid hydrolysis of the protein. The reaction is specific for glucose in ketoamine linkage and has proved reliable in most, but not all,[29,30] laboratories. Its advantages include simplicity, lack of requirement for sophisticated equipment or purchase of columns and kits, absence of interference by labile (aldimine-linked or Schiff-base) intermediates and by abnormal hemoglobins, and the ability to detect glycosylation at all glycosylated sites. The last of these features, which rests on the fact that all released carbohydrate moieties react with the TBA reagent, contrasts with ion-exchange chromatographic methods that rely on a change in the charge properties of Hb A for detection of glycosylation. For this reason, glycosylation of hemoglobin measured by the TBA reaction is usually greater than the level of glycosylation measured as Hb A_{1c} or as total "fast" hemoglobins by ion-exchange chromatography (Figure 3-1). On the other hand, results for glycosylated hemoglobin assessed by affinity chromatography on phenylboronate agarose are more directly comparable to those obtained with the TBA reaction in that the former method also detects hemoglobin modified in positions other than the NH_2-terminus of the β chain.

Disadvantges of the TBA reaction method include a low sensitivity by virtue of its critical dependence on the protein concentration and on the yield of HMF, a susceptibility to interference with free glucose,[31] the fact that the production of HMF from glycosylated hemoglobin is nonstoichiometric, and an inherent variability conferred by the use as standard of pure HMF, which does not require the oxalic acid hydrolysis step. Most of these problems can be circumvented by modifications of the described technique. For example, free glucose, which itself yields a straight-line relationship in the colorimetric assay, can be removed by dialysis or by precipitation of the protein with trichloroacetic acid (TCA) before performing the assay. Liberation of HMF can be optimized by heating to 120°C under pressure.[20,21] The use of lyophilized human hemoglobin standards with known con-

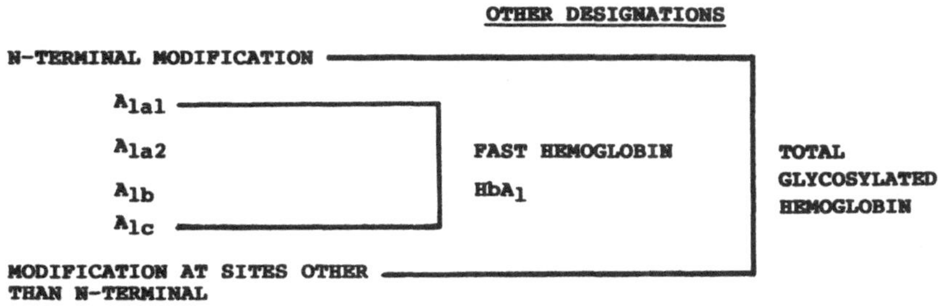

OTHER DESIGNATIONS

FIGURE 3-1   Nomenclature for glycosylated hemoglobins.

centrations of glycosylated hemoglobin addresses the concern regarding suitability of pure HMF as standard.

Specimens are collected into EDTA-containing tubes and centrifuged, and the plasma and leukocytes are removed. The packed red cells are washed twice with normal saline (0.9% NaCl) and lysed with deionized water or with a dilute solution of nonionic detergent (Nonidet P-40, 0.5% in water, v/v). After determination of hemoglobin concentration and volume adjustment to 10 g/L, a 1-mL aliquot (10 mg of hemoglobin) is placed in an acid-washed tube, to which is added 0.5 mL of oxalic acid reagent (0.9 M oxalic acid). The use of acid-washed tubes prevents contamination with carbohydrate-containing packing materials. Hydrolysis is performed for 5 hours at 100°C or for 60 minutes at 120°C in a pressure cooker or portable autoclave. The tubes are cooled in an ice-water bath, and then 0.5 mL of a solution 40% in TCA is added. The tubes are centrifuged to precipitate the protein, and 1.5 mL of the supernatant is removed to fresh tubes. After addition of 0.50 mL of the TBA reagent (0.05 M TBA), the samples are heated at 40°C for 40 minutes, and spectrophotometric absorbance is read at 443 nm against a reagent blank. The amount of HMF liberated is calculated from a standard curve obtained using 1 to 100 nmol pure HMF or from standards of human hemoglobin with known levels of glycosylation prepared from control and diabetic subjects.[20] Results are expressed as nanomoles of HMF per milligram of hemoglobin when pure HMF is used as standard, and as percent glycosylation when human hemoglobin standards are employed.

This method has also been adapted for measurement of glycosylated hemoglobin in dried blood spots,[32] which offers several theoretical advantages. Only a small amount of blood, obtainable by fingerprick rather than venipuncture, is necessary for analysis, offering features that are useful for monitoring control in children. Samples can be sent by mail to the laboratory, minimizing clinic visits for patients who live at remote distances. A small circle is marked on a glucose-oxidase-impregnated filter paper and filled with 50 μL of blood. The dried spot is eluted with 1 mL of distilled water for 1 hour and the piece of filter paper removed. From the resultant hemolysate, 50 μL is used for measurement of hemoglobin concentration, and 0.5mL of 40% TCA is added to the rest. Following centrifugation, 1 mL of 2.5 M phosphoric acid is added to the precipitate, and the samples are heated in a boiling water bath for 1 hour and then cooled. Another 0.5 mL of 40% TCA is added, the samples are centrifuged, and the supernatants are decanted to fresh tubes to which is added 0.5 mL of 0.05 M TBA. After incubation at 40°C for 30 minutes and centrifugation, spectrophotometric absorption is read at 443 nm. Reagent blank and standards (5 to 50 nm HMF) are prepared in 1 mL phosphoric acid, 0.5 mL TCA, and 0.5 ml of 0.05 M TBA, and heatd and centrifuged as are the samples.

Periodate Oxidation of Glycohemoglobin
This fluorometric assay detects the fluorescent condensation product (3,5-diacetyl-1,4-dihydrotoluidine) formed upon reaction with acetylacetone and ammonia after periodate oxidation of the carbohydrate groups in glycosylated hemoglobin.[22] Developed as a sensitive alternative to the colorimetric technique with TBA that would precisely measure the actual number of glyco groups per hemoglobin dimer, the technique has apparently not yet gained wide usage. Like the TBA procedure, this method measures glucose attaced to hemoglobin in positions other than the β chain terminus as well as in the Hb A_{1c} configuration. Since the total number of carbohydrate moieties attached to hemoglobin may provide a better reflection of diabetic control than does the percentage of Hb A_{1c}, the theoretical advantages of this method over ion-exchange chromatography are similar to those of the TBA colorimetric procedure. Additionally, reliance on imperfect liberation of HMF from the protein is avoided.

Specimens are collected into EDTA-containing tubes and centrifuged, and the packed red cells are washed three times with cold normal saline. After lysis with 5 volumes of cold distilled water and removal of red cell ghosts by centrifugation, an aliquot (1 mL) of hemolysate is added dropwise to 10mL of freshly prepared THF reagent (1% tetrahydrofuran in concentrated hydrochloric acid). The mixture is briefly agitated and centrifuged, and the precipitated globin is washed several times with cold THF that is free of acid and then dissolved in a volume of water adjusted to give a protein concentration of about 3 to 6 mg/mL. The amount of globin protein is determined by spectrophotometric absorption at 280 nm, using an extinction coefficient ($E_{1\%}$) of 8.5. Twenty microliters of 1 N HCl are added to a 700-μL aliquot of the globin preparation, followed by addition of 100 μL of 0.1 N sodium periodate. After 30 minutes at room temperature, samples are cooled on ice, and then 300 μL of 10% zinc sulfate and 100 μL of 1.4 N sodium hydroxide are added. The supernatant is collected following brief centrifugation, and 1 mL of freshly prepared formaldehyde detection reagent (2 M ammonium acetate, 0.02 M acetylacetone, in water) is added. After 1 hour at 37°C, fluorescence is measured by excitation at 410 nm with emission at 510 nm in a spectrofluorometer with a photometer attachment. The number of glyo groups per hemoglobin dimer is calculated from a standard curve obtained with fructose (0 to 40 nmol), run simultaneously with the test samples, and using the formula

$$\frac{\text{Glyco groups}}{\alpha \beta \text{ dimer}} = .032 \frac{\text{nmol fructose}}{\text{mg globin}},$$

where the molecular weight of the αβ globin dimer is taken as 32,000.

Ion-Exchange Chromatography
The separation of minor hemoglobin fractions from Hb A by ion-exchange chromatography is perhaps the most commonly employed method for measurement of Hb A_{1c}.[1,2,33-36] This method depends on the change in charge of the hemoglobin molecule produced by glycosylation of NH_2-termini of the β chains (Figure 3-2). Since Hb A_{1c} is the most abundant minor variant, and since Hb A_{1a} and Hb A_{1b} are also increased in diabetic patients,[37] measurement of the total fast fraction

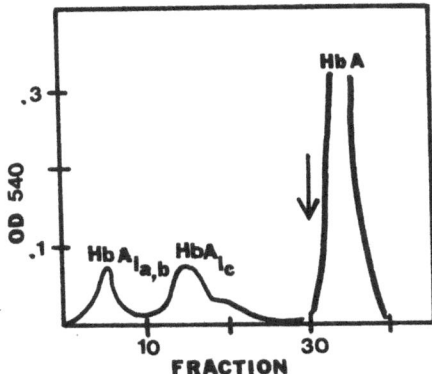

FIGURE 3-2   Elution of hemoglobins A_{1a-c} and Hb A from Bio-Rex 70 column. Arrow indicates buffer change. OD 540, optical density at 540 nm.

is adequate for assessing clinical status and glycemic control.[29,38] However, since glycosylation of α chain NH_2-termini and of ε-amino groups does not sufficiently change the charge of hemoglobin to allow chromatographic separation of these species, total glycosylated hemoglobin may be underestimated by this method.

Several small ion-exhange columns for separation of A_1 hemoglobins have been developed for commercial use and are widely available (Isolab, Inc., Akron, Ohio; Bio-Rad Laboratories, Richmond, Calif.). The manufacturers provide instruction manuals that detail the steps for assay performance, calibration, and calculation of data. Assay kits contain disposable resin columns, hemolysis reagent, elution and developing reagents, and calibrators. In recent kit editions, the hemolysis reagent contains additive to simultaneously eliminate labile Schiff-base intermediate as hemoglobin is liberated from the red blood cells during the lysis process. This is a significant addition, since it circumvents the problem of falsely elevated values due to the formation of reversible intermediates. Indeed, studies demonstrating that rapid or acute elevations in the blood glucose concentration can increase Hb A_1 levels measured by ion-exchange column chromatography initially cast some doubt on the reliability of glycosylated hemoglobin determinations by this method for evaluation of integrated diabetic control.[39,40] Exposure to increased glucose concentrations for several hours in vitro, as might occur with samples that are left unprocessed for some time after being drawn from transiently hyperglycemic patients, or in vivo, as a result of blood glucose fluctuations in daily diabetic life, promotes the formation of the Schiff-base intermediate in which glucose is reversibly linked to hemoglobin[41-46]. This labile component coelutes with irreversibly glycosylated hemoblogin on ion-exchange chromatography, but does not distort the results of the TBA procedure since it does not represent glucose in stable ketoamine linkage.[42,45,47] It does not reflect ambient blood glucose concentration over the preceding 40 to 60 days, and it is acutely responsive to changes in glucose concentration whereas the stable (ketoamine) form is not.[46-50] The labile glucoadduct contributes 2% to 3.5% to the total Hb A_1 in normal red cells and can be 7% to 9.5% of total Hb A_1 in erythrocytes from diabetic patients.[51,52] Since the labile and stable forms are indistinguishable by column chromatography, and since the former can spuriously elevate the glycosylated hemoglobin level measured by this method, it is important to remove the labile fraction before performing the assay if one wishes to monitor diabetic control reliably. This can be accomplished by incubating the erythrocytes with isotonic saline for 12 to 16 hours[51,53] or with semicarbizide and aniline for 30 minutes[54] before hemolysis. With the latter procedure, glycoadduct in the labile fraction is competitively discharged by another primary amine (semicarbazide). This is apparently accomplished simultaneously with hemolysis using commercial kits (Bio-Rad hemolysis reagent).

There are several other caveats regarding ion exchange chromatographic separation of minor hemoglobin components, such as the sensitivity of these systems to pH and temperature, the potentially confusing coelution of Hb F with glycohemoglobins, and the modest increase in Hb A_1 levels in patients with chronic renal failure and normal glucose tolerance. Variation in temperature and pH may result in values about 15% lower than those obtained with the originally described Trivelli method.[55] The manufacturers of commercial kits have largely addressed this concern by providing calibrators and graphs for temperature correction and by offering constant-temperature water baths. There is a small decline in glycohemoglobin, measured by the Isolab system, in blood from diabetic patients after a week or more of refrigerated storage.[56] Coelution of Hb F is the best known but not the only example of an abnormal hemoglobin causing an error in the determination of Hb A_1 by ion-exchange chromatography.[57-62] Since hemoglobins S, C, and D, as well as their glycosylated derivatives, co-chromatograph with Hb A, their presence will result in an underestimation of the proportion of hemoglobin that is glycosylated.[63] On the other hand, variants carrying an additional negative charge can coelute with the Hb A_1 fraction, resulting in an overestimation of the proportion of hemoglobin that is glycosylated. One such recently described hemoglobin variant gave rise to a value of 51% Hb A_1 with ion-exchange chromatography.[64] Hemoglobin variants known to elute with the Hb A_1 fraction include Hb F, Hb Hijiyama, Hb Helsinki, Hb Vassa, Hb Syracuse, and Hb Leopore.[64-68] When values for glycosylated hemoglobin determined by ion-exchange chromatography are unexpectedly high (in which case a variant co-chromatographs with the Hb A_1 fraction) or unexpectedly low (in which case the variant co-chromatographs with the Hb A fraction), the presence of an abnormal hemoglobin should be sought and an alternative method for measurement of glycosylated hemoglobin should be used.

Uremia influences the chromatographic measurement of Hb A_1 levels in both nondiabetic and diabetic patients.[69-71] The Hb A_{1a+b} portion may be higher in patients with renal failure than in diabetic patients.[72] Because uremia itself can increase Hb A_1 measured chromatographically, there is a lack of correlation between Hb A_1 and blood glucose control in uremic diabetics. The reason for this increase largely derives from the carbamylation of hemoglobin, although a shortened erythrocyte life span may also contribute. With normal glucose tolerance the proportion of total hemoglobin recovered in the

Hb A_1 fraction is 2% to 3% higher in patients with impaired renal function than in subjects with normal renal function (Figure 3-3). The species eluting with Hb A_1, on cation-exchange chromatography is the product of an interaction of hemoglobin with a urea-derived reactant, believed to be cyanate, according to the following reactions:

$$\text{a. } NH_2-\overset{\displaystyle O}{\overset{\displaystyle \|}{C}}-NH_2 \rightleftharpoons NH_4^+ + {}^-NCO$$

$$\text{b. } R-NH_2 + HN=C=O \rightarrow R-\overset{\displaystyle H}{\overset{\displaystyle \|}{N}}-\overset{\displaystyle O}{\overset{\displaystyle \|}{C}}-NH_2$$

Since uremia per se increases Hb A_1, measured chromatographically, and since glycosylated hemoglobin levels measured either chromatographically or by the TBA procedure do not correlate with blood glucose levels, assessment of glycemic control by determination of glycosylated hemoglobin with these methods cannot be used with confidence in uremic patients. A similar problem also occurs in people who abuse alcohol, since acetaldehyde, like reducing sugars

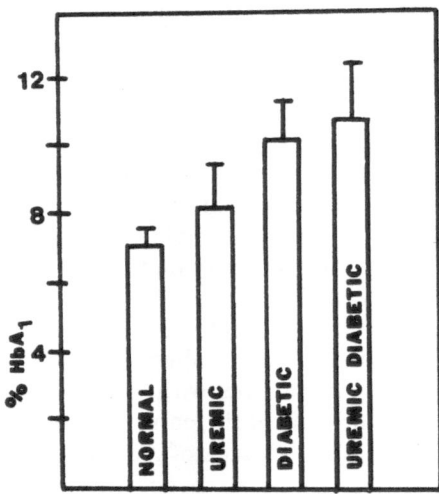

FIGURE 3-3   Hb A_1 levels in control (normal), diabetic, uremic, and uremic diabetic subjects. Adapted from Lunetta et al.[69]

and carbamate, can modify Hb A to form adducts that elute in the regions of Hb A_{1a-c} on ion-exchange chromatography.[73]

Affinity Chromatography
The affinity chromatography method is based on the ability of phenylboronic acid in alkaline solution to complex with cis-diol groups of sugars. *m*-Aminophenylboronic acid, immobilized on an agarose support matrix, specifically interacts with the 1-deoxyfructosyl derivatives of glucosylated amino acids that represent the stable forms of the products of nonenzymatic glycosylation, allowing the separation of glycosylated from nonglycosylated hemoglobin.[10,11,74] Commercial kits providing disposable minicolumns, hemolysis and elution reagents, and column regeneration solution have been developed (Glyc-Affin System, Isolab, Akron, Ohio; Glyco Gel, Pierce Chemical Co., Rockford, Ill.), and their use is becoming increasingly popular, yielding reliable and reproducible results.[7-9,12] The unbound fraction contains Hb A whereas the bound fraction, eluted with pH change or a competitive ligand such as sorbitol or other polyol, contains the glycosylated forms of hemoglobin (Figure 3-4). Because the column retains species that are glycosylated at the α chain amino terminus and the ε-amino groups of lysyl residues as well as at the β-chain NH_2-terminus (Hb A_{1c}), values for glycosylated hemoglobin determined by

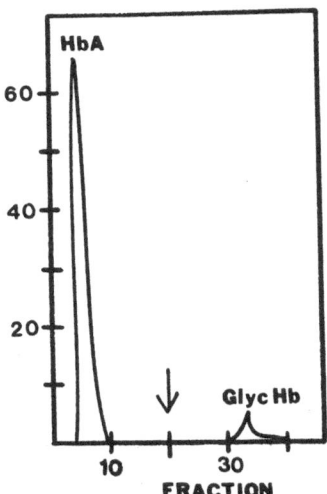

FIGURE 3-4 Elution of Hb A and glycosylated hemoglobin species from phenylboronic acid agarose column. Arrow indicates buffer change. Hb A, nonadsorbed; Glyc Hb, adsorbed at 415 nm.

affinity chromatography tend to be higher than those obtained with ion-exchange chromatography. Isoelectric focusing and HPLC analysis of the bound fraction have confirmed the presence of a heterogeneous population of hemoglobin variants, compatible with glucose modification at multiple sites. However, affinity column retention of Hb A_{1a+b} is incomplete, and hence the method probably underestimates the total number of glyco groups per hemoglobin dimer. Nevertheless, the relative insensitivity to fluctuations in pH and temperature, coupled with the lack of interference by abnormal hemoglobins such as Hb F and Hb S or by the labile aldimine, and the simplicity of performance, confer attractive advantages on this method of glycosylated hemoglobin analysis (see Table 3-1).

Isoelectric Focusing
Although column chromatography has superseded isoelectric focusing in most clinical laboratories, the latter remains a reliable though tedious technique for separation of Hb A_{1c} from Hb A. It is performed on thin-layer polyacrylamide gels, prepared by dissolving 6.79 g of acrylamide and 0.21 g of *N,N*-methylenebisacrylamide in 90 mL of distilled water to which 10 mL of ampholyte solution (4 g/10 mL water), pH 6 to 8, and 4 g of β-alanine are added.[75-78] The mixture is degassed, and then 20 μL of *N,N,N,N*-tetraethylenediamine and 400μL of freshly prepared ammonium persulfate solution (50 g/L) are sequentially added. After mixing and degassing, the solution is poured into gel molds and allowed to polymerize. The gel is mounted with a filter paper strip on a cooling plate set at 10°C, using 0.1 N sodium hydroxide as catholyte and 1 M phosphoric acid as anolyte. The gel is prefocused at a constant voltage of 0.5 kV for 30 minutes, after which samples are applied to 0.5 × 0.5-cm squares of filter paper placed at 0.5-cm intervals in a straight line in the middle of the gel. Sample size is about 2.5 μL of hemolysate in 1% KCN and should contain about 10 mg of hemoglobin. Electrofocusing is performed at a constant voltage

TABLE 3-1   Comparison of Chromatographic Methods for Measurement of Glycohemoglobin

Consideration	Ion Exchange	Affinity
Temperature sensitive	Yes	No
Affected by variant hemoglobin	Yes	No
Affected in uremia	Yes	No
Affected by labile intermediate	Yes	No
Measures glycosylation other than at *N* terminal site	No	Yes

of 0.5 kV for 30 minutes, 1.0 kV for 1 hour, and then 1.2 kV until the bands are clearly resolved. With these conditions, there is about a 2-mm separation between Hb A_{1c} and Hb A. The gel is fixed in 12.5% TCA and rinsed in glycerol solution (50 mL/L) for estimation of hemoglobin levels in a microdensitometer. The hemoglobin bands can also be cut out of unfixed gels and eluted for quantification.[79]

Agar Gel Electrophoresis

Agar gel contains negatively charged sulfate and pyruvate groups, which have different avidities for Hb A_{1c} and Hb A. The migration of Hb A is retarded, allowing a separation of about 10 mm between Hb A_{1c} and Hb A.[26,60] The agar solution is prepared by mixing agar (20 g/L), sorbitol (40 g/L), sodium citrate (35 nM), citric acid (25 mM), and Na_2-EDTA (0.9 mM) and heating to the point of solubilization of the ingredients. After pouring to form a thin film, sample wells are punched into the cooled gel. Sample size is about 1 μL (30 μg of hemoglobin), and the samples are applied to the anode side of the gel. Electrophoresis is conducted with 0.1 M citrate buffer, pH 6.3, at 60 V for 40 minutes. The gel is fixed by heat-drying at 55°C in an oven, and the amounts of Hb A and Hb A_{1c} are estimated by scanning densitometry at 420 nm. Although the resolution is good, the technique is cumbersome and agar is messy.

Serum Albumin and Plasma Proteins

Measurement of the extent of glycosylation of albumin and of plasma proteins can be performed with the colorimetric (TBA) procedure and with affinity chromatography. Both methods are now widely employed, although the commercial availability of prepackaged kits for determination by the latter method (Glyc-Affin System, Isolab) has enhanced its popularity and facilitated its use in clinical chemistry laboratories. The TBA assay has been standardized for optimal sensitivity and reproducibility in determining glycosylated serum albumin.[80] Hydrolysis at a pH of 1, achieved with 0.33 M oxalic acid, for 6 to 8 hours at 115°C is recommended for maximal product yield. Precipitation with TCA to give a 13% final concentration, and with 0.05 M TBA for 50 minutes at 37°C is recommended. In preparation for the TBA assay, albumin can be separated from freshly drawn serum by chromatography on Affi-Gel Blue (Bio-Rad) or Cibacron Blue-Sepharose (Polysciences, Inc.),[81-85] the former preceded by chromatography on DEAE-cellulose[85-87] or followed by chromatography on Bio-Gel P-150.[81] There is a reportedly excellent correlation between glucosylated albumin values determined by the TBA assay and those obtained after separation of glucosylated and nonglucosylated forms of purified albumin by chromatographic methods, where the percent glucosylated form is calculated by dividing the absorbance at 280 nm in the glucosylated peak by the sum of the absorbances in the nonglucosylated plus glucosylated peaks.[81,88] A note of caution has been introduced regarding purification of albumin on Cibacron Blue-Sepharose if sodium thiocyanate is used for elution of albumin, since this compound interferes with subsequent hydrolysis of the ketoamine and, hence, generation of HMF.[81]

The principle of separation of glycosylated from nonglycosylated forms of proteins on immobilized boronic acid is as applicable to plasma proteins and albumin as it is to hemoglobin.[14,89] After application of plasma or serum to the affinity column, nonglyco-sylated proteins elute in the alkaline wash buffer, whereas glycosylated proteins, which are adsorbed to the column, are eluted under acidic conditions. Distinction between total proteins and albumin is ac-complished by differential use of dye-reagents such as Coomassie Brillant Blue for protein and Bromocresol Green for albumin. Thus, measurement of the extent of glycosylation in total plasma proteins as well as in albumin can be obtained from a single sample applied to the affinity column, and extensive preparation of the plasma or prior separation of albumin is obviated. Quantitation is achieved by measuring the absorbance in the bound and unbound fractions at the appropriate wavelengths, and expressing the results as percentage of glycosylated protein or albumin based on the fraction bound divided by the sum of the nonbound and bound fractions. Commercial kits provide prepacked minicolumns, wash and elution buffers, standards, dye-reagents, and formulas for calculation of the percent glycosylation that take into account the elution volumes, dilution factors, and absorbance of each fraction. As shown with hemoglobin (see pre-ceding section), there is a considerable amount of TBA-reactive material present in the unbound portion of albumin or plasma protein that does not represent true glycosylated species since it does not increase with increasing levels of glycosylation and does not correlate with diabetic status. However, the bound fraction may not represent all of the glycosylated species since a small amount may escape binding to the phenylboronate column.

An alternative method for measurement of glycosylated albumin in human serum has recently been described.[90] This method is based on the observation that the fluorescent emission spectrum of dansylated phenylboronic acid (PBA) changes upon reaction with serum al-bumin, with a shift in the emission maximum from 530 to 490 nm and an increase in intensity at the latter wavelength. The presence of boric acid (H_3BO_3) in a reaction mixture of dansylated PBA plus albumin inhibits the emission intensity at 490 nm, so that calculation of the difference in the intensity of fluorescence without and with saturating amonts of H_3BO_3 provides a measure of glycosylated albumin according to the formula:

$$\frac{\Delta \text{ flourescence}}{\text{albumin (mg/mL)}} = \frac{(\text{emission intensity of A}) - (\text{emission intensity of B})}{\text{albumin (g/dL)}}$$

$$\times 10,$$

where A refers to reaction mixture containing 50 mmol/L HEPES (pH 8.5) and 0.6 µmol/L dansylated PBA, and B refers to reaction mixture containing 50 mmol/L HEPES, 0.6 µmol/L dansylated PBA, and 50 mmol/L H_3BO_3, adjusted to pH 8.5 with 1 M NaOH.

The assay is performed with 20 µL of serum that has been diluted with 4 volumes of distilled, deionized water. Simultaneous incubations are conducted for 3 to 5 minutes at 20°C in 2.0 mL total volume each of reaction mixture A and reaction mixture B. Emission intensity is read at 490 nm with excitation at 330 nm. Under these conditions, the difference in fluorescence intensities between the two reaction solutions is proportional to the percentage of glycosylated albumin (Figure 3-5). The change in fluorescence is apparently wholly ascribable to the

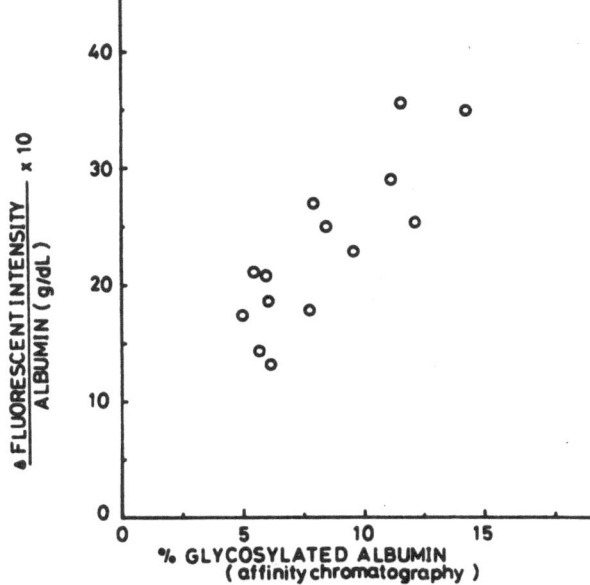

FIGURE 3-5 Relation between glycosylated albumin measured by affinity chromatography versus difference in fluorescent intensities of dansylated PBA. Reprinted with permission from Hayashi Y, Makino M: Fluorometric measurement of glycosylated albumin in human serum. *Clin Chim Acta* 1985;149:13–19. Elsevier Biomedical Press, Amsterdam.

albumin present in serum, since it disappears when serum samples are depleted of albumin.

Tissue Proteins

Thiobarbituric Acid

Measurement of glucosylated amino acid residues in tissue preparations requires hydrolysis of the polypeptide chain. Although the TBA colorimetric technique includes mild acid hydrolysis, the release of HMF from tissue proteins is often not quantitative or linearly related to protein concentration, and the presence of free carbohydrate, products of enzymatic glycosylation, or other chromogens can interfere with the accuracy of the procedure when this method is used to estimate the level of nonenzymatic glycosylation in tissue proteins. Nevertheless, the TBA reaction provides a relative measure of nonenzymatic glycosylation and has been used to demonstrate excess glycosylation of collagen and basement membranes in diabetes. Interference by free sugar is avoided by precipitating the sample with TCA, whereas borohydride reduction of the hydrolyzed samples will circumvent the problem of enzymatically glycosylated residues and other chromogens by converting aminodeoxyketose to aminohexitol, thus preventing the formation of HMF. In this instance, each sample should have its own borohydride-reduced control to serve as blank.

Borohydride Reduction

Sodium borohydride or sodium cyanoborohydride will reduce the aldimine and ketoamine products of nonenzymatic glycosylation, converting them to aminohexitols (e.g., glucitollysine, glucitolhydroxylysine). The aminohexitols can then be identified on amino acid analysis, using appropriate authentic standards prepared by chemical synthesis for calibration of elution times (see below).[91-96] Borohydride reduction is performed before acid hydrolysis in preparation for amino acid analysis since the aldimine (Schiff base) and ketoamine (Amadori rearrangement) products are labile to acid hydrolysis whereas the aminohexitol is not.[97] Reduction with tritiated borohydride is frequently used, especially when small amounts of protein are analyzed, to enhance detection of the reduced products of nonenzymatic glycosylation. Although this also results in the radiolabeling of peptide bonds that are simultaneously reduced, separation, identification, and measurement of the tritiated aminohexitols by amino acid analysis or affinity chromatography will address this problem.

Borohydride reduction is performed by mixing the protein sample with a 100-fold molar excess of sodium borohydride at alkaline pH. This can be accomplished by dissolving the borohydride in dilute sodium hydroxide (0.01 N) or sodium bicarbonate, while the protein is dissolved or homogenized in neutral phosphate buffer. The reaction is allowed to proceed for 1 to 4 hours at room temperature and then stopped by acidification to pH 4 to 5 with acetic acid. Excess borohydride is removed by exhaustive dialysis against dilute (0.05 M) acetic acid. Alternatively, excess Dowex-H+ can be added to the solution for acidification and removal of borohydride, followed by washing the resin with water and product elution with 0.5 N NH$_4$OH.[98] After evaporation several times from water to remove excess ammonia, the reduced sample is hydrolyzed in 6 N HCl in preparation for amino acid analysis, or for separation of glycosylated from nonglycosylated amino acids by affinity chromatography.

If radiolabeling of the protein is desired, reduction is performed with 10 to 25 mCi of [^{3}H]NaBH$_4$.[99] The tritiated borohydride is added quickly in several aliquots with immediate agitation and kept for 10 minutes at room temperature, followed by an additional 50 minutes at 4°C. After extensive dialysis, the borotritide-reduced protein is ready for acid hydrolysis.

Amino Acid Analysis

Acid hydrolysis is performed in 6 N HCl for 18 hours at 110°C in sealed glass ampules that have been previously evacuated or flushed with nitrogen. The acid is removed by evaporation in vacuo, and the samples are reconstituted in buffer for application to the analyzer. Several systems have been described for Beckman amino acid analyzer models and for HPLC separation, using sodium citrate buffer gradients for elution and absorbance after reaction with ninhydrin or fluorescence after postcolumn reaction with o-phthalaldehyde for detection of peaks.[92-94,97] Glucitollysine and glucitolhydroxylysine typically elute between the aromatic and basic amino acids. Since acid hydrolysis of the ketoamine produces some racemization at the second carbon atom, mannitol as well as glucitol derivatives are formed[100] (Figure 3-6) These can also be resolved on an amino-acid analyzer, as can the incompletely defined anhydroalditol derivatives.[97,101] Resolution may be imperfect, however, with pH-dependent coelution of some derivatives and confusion in the basic region of the chromatogram due to multiple anhydro derivatives.[102,103] For detection of products radiolabeled by borotritide reduction, the analyzer is equipped with a stream divider to allow portions of the effluent to be fed into a fraction collector for measurement of radioactivity. Since radiolabeled products other than hexitollysine have been identified, confirmation and

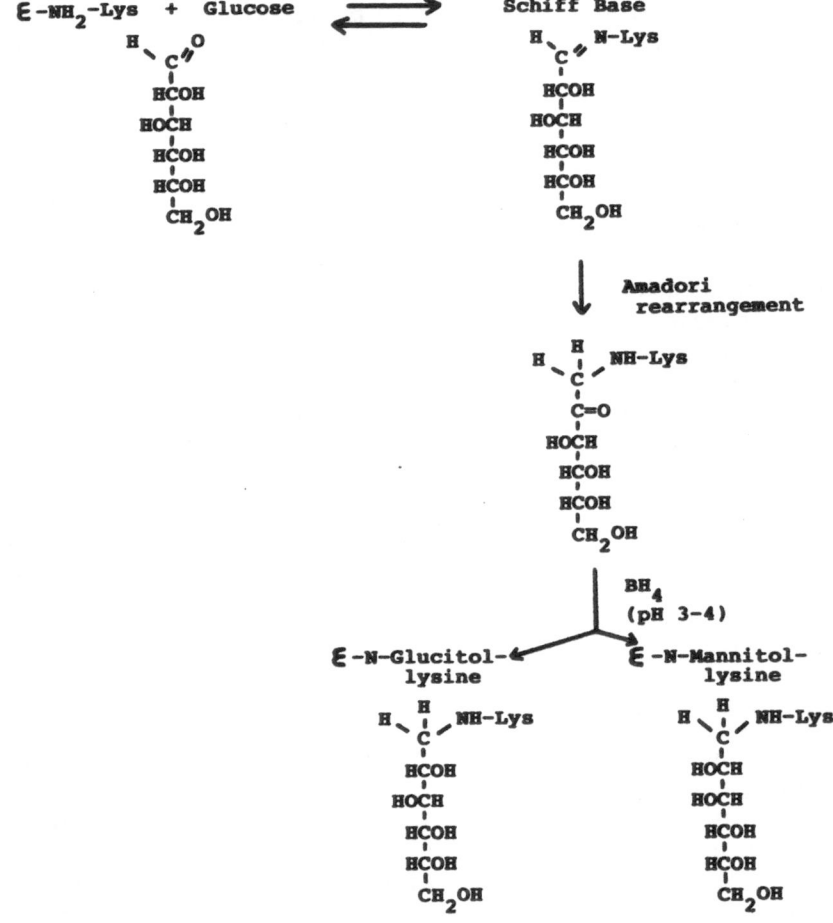

FIGURE 3-6   Racemization of Amadori product of nonenzymatic glycosylation.

quantitation of the [3H]hexitollysine may require isolation by ion-exchange or affinity chromatography.

Affinity Chromatography

Nonglycosylated and glycosylated residues of borohydride-reduced, acid-hydrolyzed proteins can be separated by application to a phenylboronate affinity column. The separating principle is the same as that utilized for measurement of glycosylated hemoglobin, namely, the ability of boronic acids to form complexes with compounds containing coplanar cis-hydroxyl groups. Samples are applied to a 0.8 × 10-cm column of immobilized PBA (Affi-Gel or PBA-60, Bio-Rad Laboratories) equilibrated with alkaline buffer (50 mM HEPES or 0.025 M sodium phosphate, pH 9.0).[74,104] After the column is washed with several bed volumes of equilibrating buffer to remove nonglycosylated amino acids and peptides, the glycosylated products are eluted with 0.1 M acetic acid or 0.025 M HCl. The effluents are monitored for ninhydrin reactivity and/or for radioactivity when reduction with tritiated borohydride has been employed. The glycosylated fraction can also be subsequently applied to an amino-acid analyzer for definitive identification of the glycosylated residues (Figure 3-7).

Furosine

The aminodeoxyketose that represents the stable ketoamine derivative after condensation of glucose with protein is an N-substituted fructosamine, 1-amino-1-deoxyfructose. Since nonenzymatic glycosylation largely involves ε-amino groups of lysine, ε-D-fructose-lysine is the principal fructosamine formed. Acid hydrolysis of this compound leads to the formation of two products, designated furosine (ε-N-[2-furoylmethyl]-L-lysine) and pyridosine (ε-[3-hydroxy-4-oxo-6-methyl-1-pyridinyl]-L norleucine, with respective yields of about 30% and 10%.[105,106] These products can be separated and detected with HPLC, allowing the development of a sensitive and specific method for the quantitative determination of the overall level of nonenzymatic glycosylation in proteins.[107] The method is based on the measurement of furosine, for which the only source in biologic materials is glycosylated amino groups of lysine residues.

In preparation for HPLC analysis, the protein sample is hydrolyzed in 6 N HCl for 18 hours at 95°C, and filtered if necessary. The HCl is evaporated under reduced pressure, and the residue is dissolved in a small amount of water. Fructose-lysine standard, prepared by chemical synthesis,[107] is subjected to the same treatment. The recommended HPLC systems are a prepacked 4 × 300-mm column of 10-μm particle size "Bondpak C_{18}" (Waters Associates, Inc.) and a 5 u C_{18}-column (4 × 200 mm). For the former, elution is performed with 7 mM H_3PO_4, and 5.6 mM H_3PO_4 is used for the elution from the 5 u C_{18} column. The latter was subsequently modified to 1.2 mM H_3PO_4 and 0.2 mM heptane sulfonic acid to achieve better resolution of furosine.[108] The effluent is monitored at 280 and 254 nm. With these conditions, furosine elutes after 3 to 3.4 minutes, followed by tyrosine at about 9 minutes and phenylalanine after about 14 minutes. Measurement of phenylalanine allows internal standardization with different proteins,

(A)

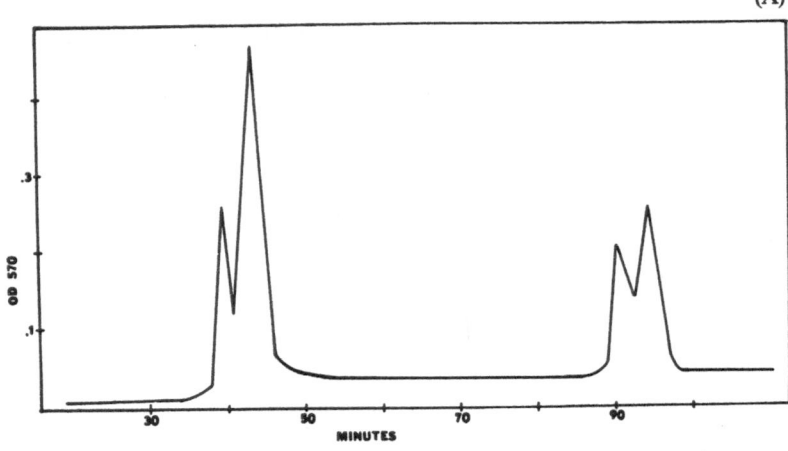

(B)

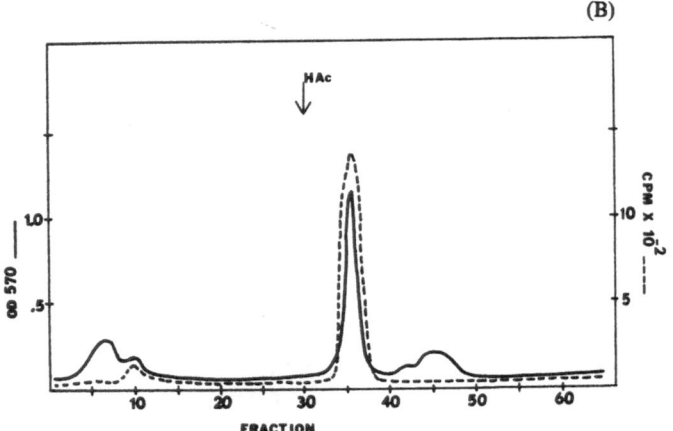

FIGURE 3-7 Purification and identification of glucitolhydroxylysine with affinity chromatography and amino-acid analysis. A, Amino-acid analysis of mixture of hydroxylysine and glucitolhydroxylysine. B, Elution of mixture of radiolabeled glucitolhydroxylysine and hydroxylysine from affinity column of agarose-linked phenylboronate. C, Amino-acid analysis of the adsorbed peak from B. All free hydroxylysine has been eliminated. OD 570, optical density at 570 nm. Reprinted with permission from *Experimental Gerontology* 18: Cohen MP, Wu V-Y: Age-related changes in nonenzymatic glycosylation of human basement membranes. Copyright 1983, Pergamon Press, Ltd.

(C)

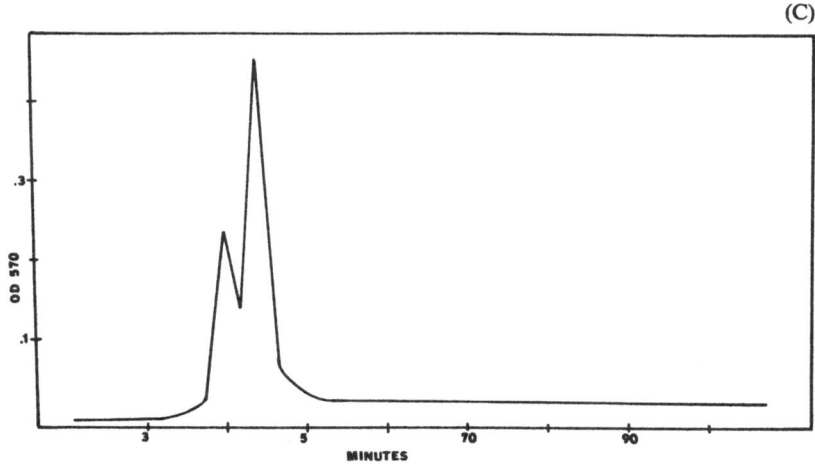

FIGURE 3.7 (*Continued*)

via calculation of furosine/tyrosine ratios. The height of the absorption peak at 280 nm shows a linear relationship with the amount of authentic fructose-lysine, providing means for calibration of the column and expression of results as nanomoles per milligram protein. Results can also be expressed as nanomoles of lysine-bound glucose (furosine) per micromole of phenylalanine or tyrosine.

References

1. Allen DW, Schroeder WA, Balog J: Observation on the chromatographic heterogeneity of normal adult and fetal hemoglobin: A study of the effect of crystallization and chromatography on the heterogeneity and isoleucine content. *J Am Chem Soc* 1958;80:1628–1634.
2. Schnek AG, Schroeder WA: The relation between the minor components of whole normal adult hemoglobin as isolated by chromatography and starch block electrophoresis. *J Am Chem Soc* 1961;83:1472–1478.
3. Clarke JT, Canivet J: Hemoglobin A_{1c} separation by microcolumn chromatography: A new rapid method of assay. *Diabete Metab* 1979;5:293–296.
4. Jones MB, Koler RD, Jones RT: Microcolumn method for the determination of hemoglobin minor fractions A_{1a+b} and A_{1c}. *Hemoglobin* 1978;2:53–58.
5. Welch SG, Boucher BJ: A rapid micro-scale method for the measurement of hemoglobin A_{1a+b+c}. *Diabetologia* 1978;14:209–211.
6. Daubresse JC, Lemy C, Bailly A, et al: The usefulness of a rapid method

for total fast hemoglobins determination in screening for diabetes control. *Diabete Metab* 1979;5:301–305.

7. Klenk DC, Hermanson GT, Krohn RI, et al: Determination of glycosylated hemoglobin by affinity chromatography: Comparison with colorimetric and ion exchange methods, and effects of common interferences. *Clin Chem* 1982;28:2088–2094.

8. Abraham EC, Perry RE, Stallings M: Application of affinity chromatography for separation and quantitation of glycosylated hemoglobin. *J Lab Clin Med* 1983;102:187–197.

9. Gould BJ, Hall PM, Cook JGH: Measurement of glycosylated hemoglobin using an affinity chromatography method. *Clin Chem Acta* 1982;125:41–48.

10. Malia AK, Hermanson GT, Krohn RI, et al: Preparation and use of a boronic acid affinity support for separation and quantitation of glycosylated hemoglobins. *Anal Lett* 1981;14:649–661.

11. Bouriotis V, Stott J, Galloway A, et al: Measurement of glycosylated hemoglobins using affinity chromatography. *Diabetologia* 1981;21:579–580.

12. Willey DG, Rosenthal MA, Caldwell S: Glycosylated hemoglobin and plasma glycoprotein assays by affinity chromatography. *Diabetologia* 1984;27:56–58.

13. Yue DK, McLennan S, Church DB, et al: The measurement of glycosylated hemoglobin in man and animals by aminophenylboronic acid affinity chromatography. *Diabetes* 1982;31:701–707.

14. Rendell M, Kao G, Mecherikunnel P, et al: Use of aminophenylboronic acid affinity chromatography to measure glycosylated albumin levels. *J Lab Clin Med* 1985;105:63–69.

15. Flückiger R, Winterhalter KH: In vitro synthesis of hemoglobin A_{1c}. *FEBS Lett* 1976;71:356–360.

16. Gabbay KH, Sosenko JM, Banuchi GA, et al: Increased glycosylation of hemoglobin A in diabetic patients. *Diabetes* 1979;28:337–340.

17. Pecoraro RE, Graf RJ, Halter JB, et al: Comparison of a colorimetric assay for glycosylated hemoglobin with ion exchange chromatography. *Diabetes* 1979;28:1120–1125.

18. Fischer W, deJong C, Voigt E, et al: The colorimetric determination of HbA_{1c} in normal and diabetic subjects. *Clin Lab Haematol* 1980;2:129–138.

19. Subramanian CV, Radhakrishnamurthy B, Berenson GS: Photometric determination of glycosylation of hemoglobin in diabetes mellitus. *Clin Chem* 1980;26:1683–1687.

20. Nicol DJ, Davis RE, Curnow DH: A simplified colorimetric method for the measurement of glycosylated hemoglobin. *Pathology* 1983;15:443–447.

21. Parker KM, England JD, DaCosta J, et al: Improved colorimetric assay for glycosylated hemoglobin. *Clin Chem* 1981;27:669–672.

22. Gallop PM, Flückiger R, Hanneken A, et al: Chemical quantitation of

hemoglobin glycosylation: Fluorometric detection of formaldehyde released upon periodate oxidation of glycoglobin. *Anal Biochem* 1981; 117:427–432.

23. Cole RA, Soeldner JS, Dunn PH, et al: A rapid method for the determination of glycosylated hemoglobins using high pressure liquid chromatography. *Metabolism* 1978;27:289–301.

24. David JE, McDonald JM, Jarret L: A high performance liquid chromatography method for hemoglobin A_{1c}. *Diabetes* 1978;27:102–107.

25. Cole RA: A new test for diabetes mellitus: The measurement of hemoglobin A_{1c} and the total fast hemoglobin using high pressure liquid chromatography. *Lab Management* 1978;16:41–44.

26. Thornton WE, Schellekens APM, Sanders GTB: Assay of glycosylated hemoglobin using agar electrophoresis. *Ann Clin Biochem* 1981;18:182–184.

27. Javid J, Pettis PK, Koenig RJ, et al: Immunologic characterization and quantification of haemoglobin A_{1c}. *Br J Haematol* 1978;38:329–337.

28. Ellis G, Diamandis EP, Giesbrecht EE, et al: An automated high pressure liquid chromatographic assay for HbA_{1c}. *Clin Chem* 1984;30:1746–1752.

29. Abraham EC, Huff TA, Cope ND, et al: Determination of the glycosylated hemoglobin (HbA_{1c}) with a new microcolumn procedure. *Diabetes* 1978;27:931–937.

30. Saibene V, Brembilla L, Bertoletti A, et al: Chromatographic and colorimetric detection of glycosylated hemoglobins: A comparative analysis of two different methods. *Clin Chim Acta* 1979;93:199–205.

31. Ma A, Naughton MA, Cameron DP: Glycosylated plasma protein: A simple method for elimination of interference by glucose in its estimation. *Clin Chim Acta* 1981;115:111–117.

32. Eross J, Kreutzmann D, Jimenez M, et al: Colorimetric measurement of glycosylated protein in whole blood, red blood cells, plasma, and dried blood. *Ann Clin Biochem* 1984;21:477–483.

33. Clegg MD, Schroeder WA: A chromatographic study of the minor components of normal adult human hemoglobin including a comparison of hemoglobin from normal and phenylketonuric individuals. *J Am Chem Soc* 1959;81:6065–6069.

34. Holmquist WR, Schroeder WA: Properties and partial characterization of adult human hemoglobin A_{1c}. *Biochim Biophys Acta* 1964;82:639–641.

35. Winterhalter KH, Glatthaar B: Chromatographic separation of HbA_{1c}. *Semin Haematol* 1971;4:84–96.

36. Chou J, Robinson CA Jr, Siegel AL: Simple methods for estimating glycosylated hemoglobins and its application to evaluation of diabetic patients. *Clin Chem* 1978;24:1708–1710.

37. Trivelli LA, Ranney HN, Lai HT: Hemoglobin components in patients with diabetes mellitus. *N Engl J Med* 1971;284:353–357.

38. Gabbay KH, Hasty K, Breslow JL, et al: Glycosylated hemoglobin and long term blood glucose control in diabetes mellitus.. *J Clin Endocrinol Metab* 1977;44:859–864.

39. Svendsen PA, Christiansen JS, Welinder B, Nerup J: Fast glycosylation of haemoglobin, letter to the editor. *Lancet* 1979;1:603.
40. Svendsen PA, Christiansen JS, Andersen RA, et al: Fast glycosylation of haemoglobin, letter to the editor. *Lancet* 1979;1:1142–1143.
41. Boden G, Master RW, Gordon SS, et al: Monitoring metabolic control in diabetic outpatients with glycosylated hemoglobin. *Ann Intern Med* 1980;92:357–360.
42. Bolli G, Compagnucci P, Cartechini MC, et al: Analysis of short-term changes in reversibly and irreversibly glycosylated haemoglobin A_1: Relevance to diabetes mellitus. *Diabetologia* 1981;21:70–72.
43. Brooks AP, Nairn IM, Baird JD: Changes in glycosylated hemoglobin after poor control in insulin dependent diabetes. *Br Med J* 1980;281:707–710.
44. Paisey R, Pennoch C, Owens C, et al: Rapid glycosylation of hemoglobin. *Diabetologia* 1981;20:80.
45. Svendsen PA, Christiansen JS, Soegaard U, et al: Rapid changes in chromatographically determined haemoglobin A_{1c} induced by short term changes in glucose concentration. *Diabetologia* 1980;19:130–136.
46. Service FJ, Fairbanks VF, Rizza RA: Effect on hemoglobin A_1 of rapid normalization of glycemia with an artificial endocrine pancreas. *Mayo Clin Proc* 1981;56:377–380.
47. Goldstein DE, Peth SB, England JD, et al: Effects of acute changes in blood glucose on HbA_{1c}. *Diabetes* 1980;29:623–628.
48. Widness JA, Rogler-Brown TL, McCormick KL, et al: Rapid fluctuations in glycohemoglobin (hemoglobin A_{1c}) related to acute changes in glucose. *J Lab Clin Med* 1980;95:386–394.
49. Botterman P: Rapid fluctuations in glycosylated haemoglobin concentration. *Diabetologia* 1981;20:159.
50. Goebel FD, Fuessel H, Dörfler H, et al: Short term changes of glycosylated haemoglobins during glucose administration in healthy and diabetic subjects. *Res Exp Med Berl* 1981;179:133–140.
51. Nathan DM: Labile glycosylated hemoglobin contributes to hemoglobin A_1 as measured by liquid chromatography or electrophoresis. *Clin Chem* 1981;27:1261–1263.
52. Huisman W, Kuijken JPAA, Tan-Tjiong HL, et al: Unstable glycosylated hemoglobin in patients with diabetes mellitus. *Clin Chim Acta* 1982; 118:303–309.
53. Ditzel J, Kjaegaard JJ, Kawahara R, et al: Glycosylated hemoglobin in relation to rapid fluctuations in blood glucose in children with insulin dependent diabetes. A comparison of methods with and without prior dialysis. *Diabetes Care* 1981;4:551–555.
54. Nathan DM, Avezzano ES, Palmer Jl: A rapid chemical means for removing labile glycohemoglobin. *Diabetes* 1981;30:700–701.
55. Arnqvist H, Cederblad G, Hermansson G, et al: A chromatographic method for measuring haemoglobin A_1: Comparison with two commercial kits. *Ann Clin Biochem* 1981;18:240–242.

56. Hamman RF, Wells R, Ryschon K, et al: Glycohemoglobin stability. *Diabetes Care* 1982;5:143–144.
57. McDonald JM, Davis JE: Glycosylated hemoglobins and diabetes mellitus. *Hum Pathol* 1979;10:279–290.
58. Aleyassine H: Glycosylation of hemoglobin S and hemoglobin C. *Clin Chem* 1980;26:526–527.
59. Aleyassine H, Gardiner RJ, Blankstein LA, et al: Agar gel electrophoresis determination of glycosylated hemoglobin: Effect of variant hemoglobins, hyperlipidemia and temperature. *Clin Chem* 1981;27:472–475.
60. Menard L, Dempsey ME, Blankstein, LA, et al: Quantitative determination of glycosylated hemoglobin A_1 by agar gel electrophoresis. *Clin Chem* 1980;26:1598–1602.
61. Bernstein RE: Glycosylated hemoglobins: Hematological considerations determine which assay for glycohemoglobin is advisable. *Clin Chem* 1980;26:174–175.
62. Aleyassine H: Low proportions of glycosylated hemoglobin associated with hemoglobin S and hemoglobin C. *Clin Chem* 1979;25:1484–1486.
63. Sosenko JM, Fluckiger R, Platt OS, et al: Glycosylation of variant hemoglobins in normal and diabetic subjects. *Diabetes Care* 1980;3:590–593.
64. Puukka R, Hekali R, Akerblom HK, et al: Haemoglobin Hijiyama: Haemoglobin variant found in connection with glycosylated haemoglobin estimation in a Finnish diabetic boy. *Clin Chim Acta* 1982;121:51–57.
65. Fitzgerald MD, Cauchi MN: Glycosylated hemoglobins in patients with a hemoglobinopathy. *Clin Chem* 1980;26:360–361.
66. Hekali R, Puuka R, Pokja R, et al: Abnormal Hb-variants can cause unexpected high values in the estimation of glycosylated hemoglobins. *J Clin Chem Clin Biochem* 1981;19:695–698.
67. Koivisto VA, Ekblom M, Icen A, et al: Abnormal hemoglobin variant: A source of error in chromatographic determination of HbA_1. *Diabetes Care* 1982;5:650–651.
68. Nicol DJ, Davis RE, McCann VJ: Hemoglobinopathy in patients with diabetes mellitus: A complicating factor in the measurement of glycosylated hemoglobin, letter to the editor. *Diabetes Care* 1983;6:524.
69. Lunetta M, Infantone E, Spanti D, et al: Glycosylated hemoglobin in uraemic patients with normal glucose tolerance or insulin dependent diabetes. *IRCS Med Sci* 1981;9:844–845.
70. deBoer MJ, Miedema K, Casparie AF: Glycosylated haemoglobin in renal failure. *Diabetologia* 1980;18:437–440.
71. Flückiger R, Harmon W, Meier W, et al: Hemoglobin carbamylation in uremia. *N Engl J Med* 1981;304:823–827.
72. Oimoni M, Yoshimura Y, Kubota S, et al: Hemoglobin A_1 properties of diabetic and uremic patients. *Diabetes Care* 1981;4:484–486.
73. Stevens VJ, Fantl WJ, Newman CB, et al: Acetaldehyde adducts with hemoglobin. *J Clin Invest* 1981;67:361–369.

74. Brownlee M, Vlassara H, Cerami A: Measurement of glycosylated amino acids and peptides from urine of diabetic patients using affinity chromatography. *Diabetes* 1980;29:1044-1047.
75. Simon M, Cuan J: Hemoglobin A$_{1c}$ by isoelectric focusing. *Clin Chem* 1982;28:9-12.
76. Flückiger R, Gallop P: Measurement of nonenzymatic protein glycosylation. *Methods Enzymol* 1984;106:77-87.
77. Spicer KM, Allen RC, Buse MG: Simplified assay of hemoglobin A$_{1c}$ in diabetic patients by use of isoelectric focusing and quantitative microdensitometry. *Diabetes* 1978;27:384-388.
78. Mortensen HB: Quantitative determination of hemoglobin A$_{1c}$ by thin-layer isoelectric focusing. *J Chromatogr* 1980;182:325-333.
79. Stickland MH, Perkins CM, Wales JK: The measurement of haemoglobin A$_{1c}$ by isoelectric focussing in diabetic patients. *Diabetologia* 1982;22:315-317.
80. Ney KA, Colley KJ, Pizzo SV: The standardization of the thiobarbituric acid assay for nonenzymatic glucosylation of human serum albumin. *Anal Biochem* 1981;118:294-300.
81. Day JF, Thornburg RW, Thorpe SR, et al: Nonenzymatic glucosylation of rat albumin. Studies in vitro and in vivo. *J Biol Chem* 1979;254:9394-9400.
82. Travis J, Bowen J., Tewksbury D, et al: Isolation of albumin from whole human plasma and fractionation of albumin-depleted plasma. *Biochem J* 1976;157:301-306.
83. Travis J, Pannell R: Selective removal of albumin from plasma by affinity chromatography. *Clin Chim Acta* 1973;49:49-52.
84. McFarland KF, Catalano EW, Day, JF, et al: Nonenzymatic glucosylation of serum proteins in diabetes mellitus. *Diabetes* 1979;28:1011-1013.
85. Dolhofer R, Renner R, Wieland OH: Different behavior of haemoglobin A$_{1a-c}$ and glycosyl-albumin levels during recovery from diabetic ketoacidosis and non-acidotic coma. *Diabetologia* 1981;21:211-215.
86. Dolhofer R, Wieland OH: Glycosylation of serum albumin: Elevated glycosyl-albumin in diabetic patients. *FEBS Lett* 1979;103:282-286.
87. Dolhofer R, Wieland OH: Increased glycosylation of serum albumin in diabetes mellitus. *Diabetes* 1980;29:417-422.
88. Day JF, Thorpe SR, Baynes JW: Nonenzymatically glucosylated albumin. *J Biol Chem* 1979;254:595-597.
89. Willey DG, Rosenthal MA, Caldwell S: Glycosylated haemoglobin and plasma glycoprotein assays by affinity chromatography. *Diabetologia* 1984; 27:56-58.
90. Hayashi Y, Makino M: Fluorometric measurement of glycosylated albumin in human serum. *Clin Chim Acta* 1985;149:13-19.
91. Bailey AJ, Robins SP, Tanner MJA: Reducible components in the proteins of human erythrocyte membrane. *Biochim Biophys Acta* 1976; 434:51-57.

92. Houesly TJ, Tanzer ML: The separation and amino acid analysis of collagen crosslinks on an extended basic ion-exchange column. *Anal Biochem* 1981; 114:310–315.

93. Cohen MP, Wu V-Y: Identification of specific amino acids in diabetic glomerular basement membrane collagen subject to nonenzymatic glucosylation in vivo. *Biochem Biophys Res Commun* 1981;100:1549–1554.

94. Wu V-Y, Cohen MP: Reducible cross-links in human glomerular basement membrane. *Biochem Biophys Res Commun* 1982;104:911–915.

95. Walton DJ, Ison ER, Szarek WA: Synthesis of N-(1-deoxyhexitol-1-yl) amino acids, reference compounds for the nonenzymic glycosylation of proteins. *Carbohydr Res* 1984;128:37–49.

96. Trueb B, Hughes GJ, Winterhalter KH: Synthesis and quantitation of glucitollysine, a glycosylated amino acid elevated in proteins from diabetics. *Anal Biochem* 1982;119:330–334.

97. Baynes JW, Thorpe SR, Murtiashaw MH: Nonenzymatic glucosylation of lysine residues in albumin. *Methods Enzymol* 1984;106:88–98.

98. Schwartz B, Gray GR: Proteins containing reductively aminated disaccharides. *Arch Biochem Biophys* 1977;181:542–549.

99. Bookchin RM, Gallop PM: Structure of hemoglobin A_{1c}: Nature of the N-terminal βchain blocking group. *Biochem Biophys Res Commun* 1968;32:86–93.

100. Bunn HF, Gabbay KH, Gallop PM: The glycosylation of hemoglobin: Relevance to diabetes mellitus. *Science* 1978;200:21–27.

101. Robins SP, Bailey AJ: Age-related changes in collagen: The identification of reducible lysine-carbohydrate condensation products. *Biochem Biophys Res Commun* 1972;48:76–84.

102. Rucklidge GJ, Bates GP, Robins SP: Preparation and analysis of the products of non-enzymatic protein glycosylation and their relationship to cross-linking of proteins. *Biochim Biophys Acta* 1983;747:165–170.

103. Kennedy L, Baynes JW: Non-enzymatic glycosylation and the chronic complications of diabetes. *Diabetologia* 1984;26:93–98.

104. Cohen MP, Wu V-Y: Age-related changes in nonenzymatic glycosylation of human basement membranes. *Exp Gerontol* 1983;18:461–469.

105. Finot PA, Bricout J, Viani R, et al: Identification of a new lysine derivative obtained upon acid hydrolysis of heated milk. *Experientia* 1968;24:1097–1099.

106. Finot PA, Viani R, Bricout J, et al: Detection and identification of pyridosine, a second lysine derivative obtained upon acid hydrolysis of heated milk. *Experientia* 1969;25:134–135.

107. Schleicher E, Wieland OH: Specific quantitation by HPLC of protein (lysine) bound glucose in human serum albumin and other glycosylated proteins. *J Clin Chem Clin Biochem* 1981;19:81–87.

108. Vogt BW, Schleicher ED, Wieland OH: ε-amino-lysine-bound glucose in human tissues obtained at autopsy: Increase in diabetes mellitus. *Diabetes* 1982;31:1123–1127.

CHAPTER 4

Clinical Use

Monitoring Glycemic Control

Glycohemoglobin

There is no question that periodic monitoring of glycosylated hemoglobin is useful for documenting the degree of glucose control that has prevailed during an interval of several weeks before the sample is taken, since the glycohemoglobin concentration reflects the time-averaged concentration of glucose within the erythrocyte during that period. Evidence supporting the correlation between glycohemoglobin levels and traditional methods of assessing glucose control in insulin-dependent diabetic patients is abundant.[1-5] Correlations are best when multiple glucose measurements over several weeks are analyzed, and are not found when a single blood glucose or 24-hour urine glucose determination is used. A study of 18 counselors with insulin-dependent diabetes attending an 8-week camp session found that the percent sugar-free urine test was the best predictor of normal or elevated Hb A_{1c} values, although there was also a significant correlation between Hb A_{1c} and mean preprandial blood glucose concentrations.[6] Two earlier studies had also found that Hb A_{1c} correlated highly with glycosuria measured three times per day with Clinitest detection during the preceding 8 weeks.[7,8] In essence, glycosylated hemoglobin correlates with long-term metabolic control, regardless of whether such control is achieved by multiple or single insulin injections, insulin-pump therapy, or residual beta cell function.[9-12] Thus, periodic measurement of glycohemoglobin levels provides an objective assessment of glycemic status that complements and extends information gained from traditional methods to evaluate glucose control and treatment regimens. In this context, the technique emloyed to measure glycosylated hemoglobin is less critical than consistent use of the same technique and its reliable performance since, as discussed in earlier chapters, normal ranges will differ according to what is being measured by different methods.

The use of glycosylated hemoglobin or Hb A_{1c} determinations as a parameter of glycemic control in diabetic patients has gained widespread popularity. Indeed, advertisements appearing in the medical and even the lay press exhorting physicians and patients to avail themselves of the benefits gleaned from glycohemoglobin measurements have created the impression that failure to do so borders on negligence. There is little doubt that, when properly performed both technically and temporally, glycosylated hemoglobin levels can provide valuable information of practical, clinical relevance to the anti-diabetic regimen. This is best appreciated in situations where blood glucose concentrations are in the normal range and the patient is relatively aglycosuric. In such instances, measurement of glycosylated hemoglobin will allow the physician to assess whether these glucose values, determined intermittently during office visits or more regularly with home glucose monitoring, truly reflect euglycemia over an integrated period of time. Sporadic fasting and random blood glucose levels may be relatively normal despit wide swings in blood glucose and/or an increased mean glycemic excursion.[7,8] Similarly, some patients will make conscious, overt efforts to follow their antidiabetic program rigidly during the few days preceding a visit to their physician. This short-term compliance is not necessarily representative of their daily routine. Others may fabricate or record values for home glucose determinations that were never performed, or otherwise do not represent true values, perhaps because they need to deceive themselves or because they want to please the physician.[13,14] In fact, discrepancies between the degree of control indicated by glycosylated hemoglobin determinations versus that reflected in records of self-monitored urine or blood glucose levels should alert the physician to these possibilities. A recent report described 10 underlying psychological mechanisms that might prompt falsification of monitoring records, and suggested that lack of concordance between glycohemoglobin and glucose values could be used to detect psychological problems.[14] These include attempts to seek perfection, approval, or independence; to avoid punishment, criticism, or depression; and to express denial, anger, or guilt. Although these problems require guidance and counseling for their resolution, the fact that such deceptions can be identified and unmasked with measurement of glycosylated hemoglobin may be helpful in raising patient awareness and in the adjustment process. One group even taught the procedure for measuring glycosylated hemoglobin to some of their patients, and suggested that self-monitoring of Hb A_{1c} (as well as blood glucose) can provide additional incentive and motivation to diabetic patients that will have a positive impact on overall diabetic control.[15]

There is little point in monitoring a patient's diabetic control with repeated glycosylated hemoglobin determinations if the blood and/or urine glucose values are continuously elevated. On the other hand, in patients with apparent good control as assessed by blood glucose concentrations, and who are on a stable therapeutic regimen, one glycohemoglobin determination in a 3-month period appears adequate.[16] Further, there is a good correlation between single, random measurements of morning blood glucose or quantitative glucose excretion and Hb A_{1c} levels in stable, insulin-independent diabetics.[8] This relationship is disturbed if therapy is interrupted. For example, in a study of 25 well-controlled, sulfonylurea-treated patients, fasting plasma glucose concentrations rose from 128 ± 6 mg/dl within 2 weeks after therapy was discontinued, but an increase in Hb A_{1c} was demonstrable only after 4 to 6 weeks had elapsed.[17] Institution of therapy designed to lower fasting plasma glucose concentrations to less than 6 mmol/L in non-insulin-dependent diabetes will reduce Hb $A_{1c_1}c$ to the high-normal range.[18]

The utility of glycohemoglobin measurements in assessing responses to changes in therapy in non-insulin-dependent diabetes reportedly can be increased by application of a calculated formula.[19] The approach is based on the observation that there is a highly significant linear relationship between the fasting plasma glucose concentration and the glycosylated hemoglobin value in patients with type II diabetes, which allows graphically determined estimates of equilibrium values for glycosylated hemoglobin that would correspond to measured fasting plasma glucose levels.[20] The formula offers a parameter designated the glycosylated hemoglobin index (GHbI) as follows:

$$GHbI = \frac{[GHb] \text{ expected}}{[GHb] \text{ actual}},$$

where the expected value is the equilibrium value derived as described above. This approach supposedly allows fasting glucose and glycohemoglobin concentrations obtained on a single occasion to give information regarding changes in glycemic control that occurred over intervals of 1 to 4 weeks before the samples were obtained, and thus provides a means of assessing recent improvement or deterioration after changes in therapy.

Glycoalbumin and Glycosylated Serum Proteins

The nonenzymatic glycosylation of serum albumin is increased in patients with poorly controlled diabetes[21-23] and in rats with alloxan-induced diabetes.[24,25] In one of the latter studies, a four fold elevation of glycosylated albumin occuring 4 days after withdrawal of insulin suggested that its measurement could be used as a sensitive indicator of recent glycemic control. Indeed, the initial description of the in vitro and in vivo nonenzymatic glucosylation of rat and human serum albumin was quickly followed by the report of several patients with increased levels of glycosylated albumin despite normal levels of glycosylated hemoglobin.[23,24,26] This was consistent with the concept that nonenzymatic glycosylation of albumin, a protein with a shorter circulating half-life than the red blood cell, reflects blood glucose concentrations over a shorter, more recent (1 to 2 weeks) period than does Hb A_{1c}. The temporal relationship between changes in blood glucose concentrations and the level of glycosylated albumin was further elucidated in a study of 12 diabetic patients recovering from diabetic ketoacidosis or nonketotic hyperglycemic (hyperosmolar) syndrome.[27] At the time of the acute hyperglycemic episodes, average initial values for percent Hb A_1 and for glycoalbumin (in nanomoles of HMF per mole of albumin) were increased about three-fold compared to normal values. During the first week of therapy, changes in Hb A_1 were negligible, whereas the level of glucosylated albumin diminished to 87% of initial values. During the subsequent 10 days, glucosylated albumin continued to decline, reaching levels about 50% those of initial values by day 17, whereas only a small change in Hb A_1 was observed by the end of the same interval (Figure 4-1). In a recent study of 73 children attending a summer camp for children with diabetes, mean initial glycosylated albumin levels fell from 16.4% to 14.6% after 10 days of careful control, and the final level of glyco-albumin correlated with fasting blood glucose concentrations in reflecting improved control.[28] Mean glycosylated serum transferrin fell from 8.2% to 7.0% during this same period in these children.

Although albumin is quantitatively the most important glucosylated plasma component, plasma proteins other than albumin also undergo glycosylation, and their measurement is a useful index of diabetic control.[29-31] The extent of serum protein glycosylation correlates with

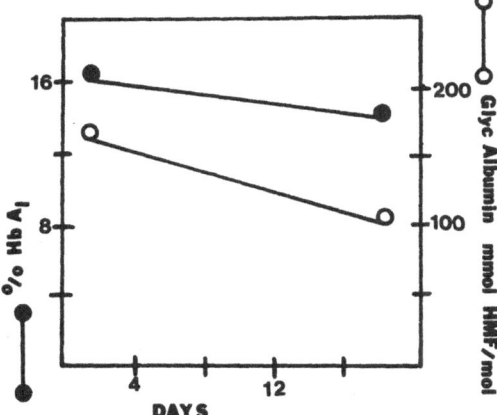

FIGURE 4-1   Response of glycosylated hemoglobin and glycosylated albumin to correction of severe hyperglycemia over a 17-day period. Adapted from Dolhofer et al.[27]

fasting blood glucose concentrations and with percentage of glycosylated hemoglobin in diabetic patients, and agreement between values for glycosylated albumin and glycosylated serum proteins is excellent in specimens from both normal and diabetic subjects. Further, in contrast to Hb A₁, there is no overlap in values for glycosylated serum proteins between normal and diabetic populations.[29] Although at times the level of glycosylated serum proteins may not correspond to a single blood glucose determination,[32] it does correlate with the degree of diabetic control estimated by the mean and variance of multiply determined plasma glucose concentrations,[30] or by mean glucose values judged from four Chemstrip readings per day.[33] Measurement of glycosylated serum or plasma protein, like glycosylated albumin, may provide a better index of diabetic control than Hb A₁c in patients with hemolytic anemia, hemoglobinopathy, or recent blood transfusion. Since the half-life of many circulating proteins is only a few days, and since much of the glycosylation of serum proteins is attributable to albumin, which also has a relatively short circulating half-life, changes in the level of serum protein glycosylation in response to changes in glycemic status occur more rapidly than do changes in glycosylated hemoglobin. Thus, glycosylated serum protein reflects the average blood glucose concentration during the 15-day interval before sampling. For reliable results that accurately reflect glycemic status, specimens must be dialyzed to remove free glucose if the colorimetric (thiobarbituric acid) reaction is employed.[34] However, measurement of glycosylated proteins or albumin by affinity chromatography is rapidly replacing the color-imetric procedure. This method is less cumbersome, avoids dialysis or other preparation of the plasma, and appears more reliable than the TBA reaction procedure.[35,36] Analysis of the unbound (nonglyco-sylated) and bound (glycosylated) fractions after application of undialyzed plasma samples to aminophenylboronic acid columns has revealed that the bulk of TBA-reactive material is present in the former, and probably represents free glucose. This fraction does not correlate with diabetic status or with the extent of nonenzymatic glycosylation. In contrast, TBA reactivity of the bound fraction correlates with glycemic status and with the extent of nonenzymatic glycosylation determined by glycohemoglobin values.

The clinical value of glycosylated serum protein or albumin determinations, especially when measured by affinity chromatogra-phy, is becoming increasingly appreciated because of their ability to provide an index of glycemic status during the preceding 1 to 4 weeks.[37] Information obtained from glycosylated albumin levels complements, but does not replace, monitoring of long-term control by measurement of glycohemoglobin levels. The real niche for measurement of glycosylated albumin lies in the ability of the results to provide an objective index of response to changes in the therapeutic regimen that is temporally more immediate than that reflected in glycohemoglobin levels.

An interesting caveat concerning nonenzymatic glucosylation of serum albumin that has potential clinical significance has recently been reported.[38] This study examined the impact of glycosylation on the sulfonylurea-binding capacity of albumin, to which these drugs normally bind strongly. Glucosylation of serum albumin resulted in a decrease to about 50% of control values in the amount of tolazamide, acetohexamide, glibenclamide, and tolbutamide bound to this protein. Since decreased binding would give rise to an increase in the amount of free drug in the circulation, the clinical import of this finding is obvious. The tissue availability and utilization of sulfonylureas may be altered in patients with high levels of glycosylated albumin, and the potential for adverse effects of these agents might be increased in such patients. The change in sulfonylurea-binding capacity may be repre-sentative of a more general effect on the ligand-binding properties of albumin, since the affinities of bilirubin and of long-chain fatty acids for glycosylated albumin are reduced compared to those for the nonglycosylated forms.[39] Whether these effects on ligand binding translate to clinically significant correlates is currently unknown.

Although measurement is feasible, the amount of glycosylated amino acids or peptides excreted in the urine is apparently not a useful parameter for assessment of metabolic control.[40,41] The level of urinary glycosylated amino acids, measured by affinity chromatography of borohydride-reduced acid hydrolysates, correlates with body weight and is increased in the urine of diabetic subjects. However, the daily excretion of glycosylated peptides, measured as nanomoles HMF per 24 hours, does not correlate with mean daily blood glucose levels and, in patients with proteinuria, does not correlate with glycosylated serum proteins or with glycohemoglobin values.

Diagnosis

The notion that a simple blood test using a single sample obtained at any time of day could replace a time-consuming and unpleasant procedure, requiring dietary preparation and multiple blood samples and yielding results that are interpreted according to variable criteria, is attractive. However, initial speculation that measurement of glycosylated hemoglobin could be employed as a more sensitive or specific diagnostic test than the standard oral glucose tolerance test (GTT) has not been entirely corroborated in clinical studies.[43–46] The glycosylated hemoglobin level does correlate with blood glucose concentrations observed during the GTT, and there is a direct correlation with the area under the curve in diabetic patients.[3] However, the strongest correlation exists between Hb A_{1c} and fasting glucose levels.[43,44–46] Since response to oral glucose challenge is not necessary to document the diagnosis of diabetes in the presence of overt (fasting) hyperglycemia, the demonstration that Hb A_{1c} is elevated in such patients reflects its established ability to estimate integrated blood glucose levels rather than an ability to discriminate between diabetic and nondiabetic populations. Nevertheless, the combination of an elevated fasting blood glucose concentration and an increased glycosylated hemoglobin level obviates the need for further testing and indicates clinically significant hyperglycemia. The results of a recent screening survey suggest that the relative sensitivity of Hb A_{1c} is slightly better, although the relative specificity is slightly less, than fasting plasma glucose determinations for the detection of diabetes, and that the best predictive value for a positive diagnosis can be obtained from a combination of these two tests.[47]

Although some studies have found a correlation between glycosylated hemoglobin and peak glucose values in patients with impaired glucose tolerance but normal fasting blood glucose concentrations (<140 mg/dL), Hb A_{1c} is actually elevated only in patients whose 2-hour glucose levels are above 228 mg/dL (Table 4-1). From this perspective, Hb A_{1c} measurement offers no added advantage over

TABLE 4-1   Range of Hb A$_1$ Values According to Glucose Tolerance*

Hb A$_1$ (%)	Plasma Glucose (mg/dL)		
	Fasting	1 Hours	2 Hours
6.5–9.9	<112	<200	<140
6.2–10.7	<140	180–280	<200
6.9–11.2	<140	200–390	210–350
7.8–16.2	>140	260–515	240–525

*Data from Lev-Ran.[42]

blood glucose values obtained with standard glucose tolerance testing.[42,43,48-50] An alternative view might hold that criteria for the diagnosis of glucose intolerance should include an elevated glycohemoglobin level[51]; in other words, a more conservative approach to the diagnosis of diabetes should be taken. This interpretation is suggested by the finding that Hb A$_1$ levels were not significantly different from control in a group of individuals with glucose intolerance according to the Fajans-Conn criteria (peak > 185 mg/dL), whereas glycohemoglobin values were elevated in 86% of subjects judged glucose intolerant by 1- and 2-hour values exceeding limits of 259 and 219 mg/dL, respectively.[46]

The mean Hb A$_1$ in 67 individuals with abnormal oral GTTs was significantly greater than that in 40 normal subjects (7.7% ± 0.09% versus 6.4% ± 0.08%) (Figure 4-2). However, the range of values in each group overlaped (6.3% to 9.6% versus 5.2% to 7.2% in subjects with abnormal versus normal GTTs), and 14 people in the group with abnormal glucose tolerance had Hb A$_1$ levels less than 7.2%.[52] Findings were similar in another study of 37 normal and 21 borderline (glucose > 123 mg/dL 2 hours after a 100-g oral glucose load) diabetic subjects.[53] Of the latter group, only 11 individuals had glycohemoglobin values above the normal range. A recent survey of 819 subjects between the ages of 47 and 54 years found that the mean Hb A$_1$ was 7.3% in 51 people with impaired glucose tolerance after a 75-g oral carbohydrate challenge, compared to a mean value of 6.5% in 150 subjects with normal glucose tolerance.[54] Of the glucose-intolerant group, only 20% were considered to have an elevated Hb A$_1$ when the upper limit of normal (mean ± 2 standard deviations) was taken as 7.8%. There were no significant differences with respect to fasting and 2-hour blood glucose concentrations or to the area under the curve between glucose-intolerant patients with lower (mean = 6.9%) versus higher (mean = 7.9%) levels of Hb A$_1$. As in other studies, Hb A$_1$ strongly correlated with fasting glucose and with postchallenge glucose

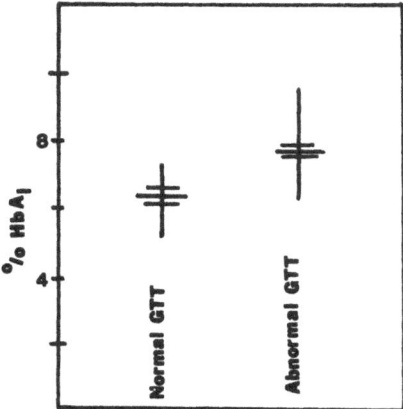

FIGURE 4-2   Hb A₁ in subjects with normal and abnormal GTTs. Horizontal lines represent mean ± SEM; vertical lines represent range of values in each group. Adapted from Bolli et al.[52]

levels and the area under the curve after a 75-g oral glucose load in patients with non-insulin-dependent diabetes. Thus, although Hb A₁ levels can corroborate the presence of manifest diabetes, their measurement appears to be of little value as a screening test for detection of diabetes in view of the overlap in the range of values found in normal and mildly glucose-intolerant populations. This experience is similar to that reported in pregnant women, as discussed in the next section. It is possible that determination of glycosylated hemoglobin with affinity chromatography, rather than with ion-exchange chromatography, could provide a more clear-cut separation between normal and glucose-intolerant subjects,[55] but this awaits confirmation. The level of glycosylated hemoglobin does increase slightly with age, a finding probably related to the modest deterioration in carbohydrate tolerance that occurs with aging, but this also cannot be used diagnostically.[56] There may also be a sex difference in that Hb A₁ levels are reportedly higher in men than in women.[57]

One situation in which measurement of glycosylated hemoglobin may prove of diagnostic use is in patients with myocardial infarction or other stressful illness in whom hyperglycemia is detected in the coronary care unit or during hospitalization.[58] Although hyperglycemia is frequently encountered following acute myocardial infarction, an elevated Hb A₁ occurs less commonly and is seen equally in patients with and without myocardial infarction. Thus, measurement of glycosylated hemoglobin in patients with acute intercurrent illness and hyperglycemia may distinguish transient hyperglycemia secondary to stress from that due to diabetes mellitus.

Recently, estimation of the extent of nonenzymatic glycosylation of serum proteins by HPLC measurement of furosine (see Chapter 3) has been able to distinguish some patients with abnormal oral GTTs from those with perfectly a normal response to oral glucose challenge.[59] Of 15 patients given 75 g of carbohydrate, the 8 with completely normal glucose values had normal levels of serum lysine-bound glucose, whereas 25% of the 7 patients with abnormal glucose responses also had elevated lysine-bound glucose levels. Of 45 patients given 100 g of glucose, 27 had normal glucose and lysine-bound glucose values, whereas one fifth of the abnormal glucose responders also had elevated lysine-bound glucose levels. Measurement of serum furosine appears to be a method suitable for monitoring even small fluctuations of blood glucose, since two patients with insulinoma reportedly had diminished values.

Pregnancy

In nondiabetic women, Hb A_{1c} levels progressively decrease from the first to the third trimester of pregnancy,[60-62] presumably because of lower mean blood glucose concentrations.[63] Levels also decrease during the third trimester in diabetic women, perhaps in part due to improved control.[60,64-66] This decrease is relative to values in the first trimester or to levels in nonpregnant counterparts[67]; values are actually increased compared to those observed at each trimester interval in normal women (Figure 4-3).

As in the nonpregnant patient, an elevated Hb A_{1c} during pregnancy reflects inadequate diabetic control and indicates that the mother was hyperglycemic during the preceding 4 to 6 weeks or more. Repeatedly elevated values reflect chronic poor control. In either case, an adequate antidiabetic regimen must be instituted. With poor control, excess metabolic fuels cross the placental barrier and promote fetal insulin secretion.[68-70] Indeed, the characteristic macrosomia in infants of diabetic mothers appears, to a large extent, to arise from fetal hyperinsulinemia provoked by hyperglycemia and by exposure to increased amounts of other insulin secretagogues.[71-75] It is thus not surprising that birth weight ratios (birth weight divided by 50th percentile for sex-corrected expected birth weight for gestational age) are higher in infants born to diabetic mothers with elevated Hb A_{1c}, and that there is a significant correlation between birth weight ratio and maternal Hb A_{1c}.[74,76] Some[74,76,77] but not all[66,78-80] studies suggest that even one elevated value can be used to predict infant birth weight

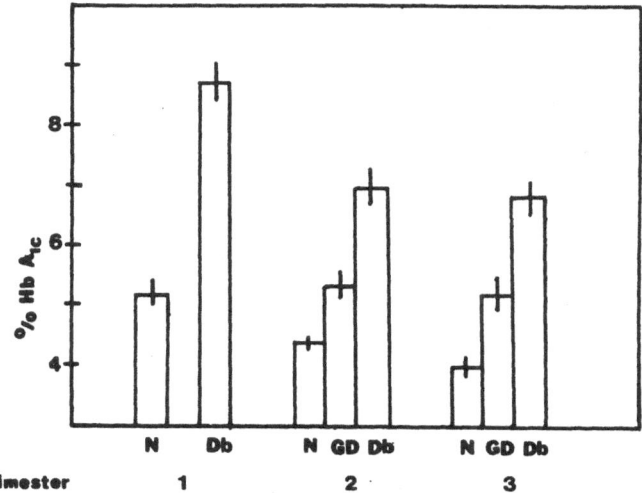

FIGURE 4-3 Hb A$_{1c}$ levels during pregnancy in nondiabetic (N), gestational diabetic (GD), and diabetic (Db) women. Adapted from Widness et al.[60]

(Figure 4-4). Maternal status apparently influences the predictive value, since the best correlation between Hb A$_{1c}$ and relative birth weight ratio is found when only results from women with White classes B and C without prognostically bad signs in pregnancy are analyzed.[80] The time of gestation at which the blood sample is drawn can affect the accuracy of prediction. Since the influence of hyperglycemia on fetal somatic growth appears to be greatest between the 26th and 32nd weeks of gestation, an elevated Hb A$_{1c}$ at 32 weeks is probably a more likely predictor of birth weight than one at 40 weeks.[81] An infant may be large at birth even if the mother's Hb A$_{1c}$ level is normal at 38 or 40 weeks and control was optimum during the interval subsequent to 32 weeks, if hyperglycemia had prevailed during the preceding 4 to 6 weeks. On the other hand, poor control in the early weeks of pregnancy may inhibit fetal growth, since elevated Hb A$_{1c}$ in the first trimester has been associated with small fetuses.[82] High levels of Hb A$_{1c}$ in early pregnancy have been associated with congenital anomalies.[65,83,84] Increased maternal glycohemoglobin levels at delivery tend to be associated with high umbilical cord levels of C-peptide, reflecting fetal hyperinsulinemia. Glycosylated hemoglobin in cord blood is higher in hypoglycemic infants of diabetic mothers than in infants who do not develop hypoglycemia.[79,85] Thus, neonatal hypoglycemia results from fetal hyperinsulinemia consequent to maternal hyperglycemia.

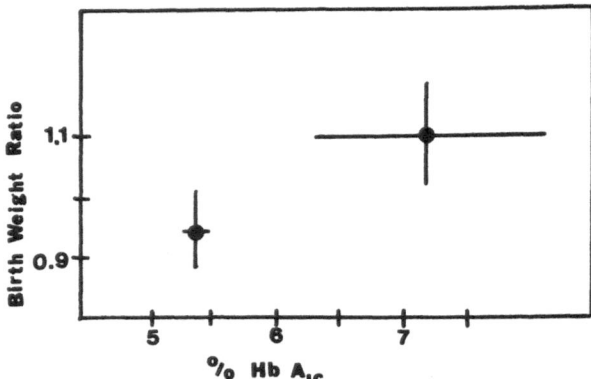

FIGURE 4-4 Maternal Hb A$_{1c}$ levels and infant birth weight ratio. Adapted from Susa et al.[74] and Susa and Schwartz.[75]

Most attempts to use Hb A$_{1c}$ levels during pregnancy for diagnostic or screening purposes to detect gestational diabetes have not met with success.[60,66,86,87] The mean level in a group of women with gestational diabetes may be significantly greater than in a group of nondiabetic counterparts, but the range of values in each population generally overlaps to an extent that precludes diagnostic utility on an individual basis. For example, in one study the mean Hb A$_{1c}$ in six pregnant women with normal GTTs was 6.3% with a standard error of 0.7% and a range of 5.5% to 7.3%. The mean Hb A$_{1c}$ in eight pregnant patients who had abnormal results after oral glucose challenge was 6.2% with a standard error of 1.1% and a range of 4.6% to 8.1%.[66] The mean Hb A$_{1c}$ level in another group of 13 normal pregnant women during the second trimester was 4.4%, compared to 5.2% in 12 women with carbohydrate intolerance and a similar gestational period. However, the range of values in the nondiabetic pregnant women was 3.9% to 6.8% during the second trimester.[60] Another study, which used the TBA acid reaction and expressed data as nanomoles of hydroxymethyl-furfuraldehyde per milligram of protein, found no difference in glycosylated hemoglobin when levels in 41 women with normal glucose tolerance were compared with those in 12 women with gestational diabetes mellitus.[86] Samples were obtained between the 24th and the 32nd week of gestation, and the diagnosis of gestational diabetes was made if two or more blood glucose values at timed intervals after 100 g of oral glucose exceeded 190 mg/dL at 1 hour, 165 mg/dL at 2 hours, or 145 mg/dL at 3 hours. On the other hand, the mean value for glycosylated serum protein in eight of these gestational diabetic subjects was significantly greater than that in 17 of the women with normal glucose tolerance (0.54 ± 0.06 [SD] versus 0.49 ± 0.07 nmol hydroxymethylfurfuraldehyde per mg of protein). However, there was sufficient overlap in the range of values observed in samples from these two groups of patients to preclude the use of a single glycosylated serum protein level to categorize a woman as having gestational diabetes. Thus, the sensitivity of the oral GTT for detection of carbohydrate intolerance during pregnancy is superior to that achieved with glycosylated protein or glycohemoglobin measurements. As discussed in the preceding section, the experience is similar in nonpregnant populations, wherein measurement of glycosylated hemoglobin appears to be of little value as a screening test for detection of diabetes.

Complications of Diabetes

Although the general consensus now subscribes to the view that sustained hyperglycemia contributes to the pathogenesis of several complications of diabetes, glycosylated hemoglobin levels do not necessarily predict, nor correlate with the presence of, various complications. The reason for this probably relates, at least in part, to the fact that many studies used only one glycohemoglobin determination, rather than multiple and sequential measurements, to examine these relationships. A single glycosylated hemoglobin level provides an integrated measurement of blood glucose control that pertains only to the 4- to 8-week period preceding the determination. Long-term prospective clinical studies, in which Hb A_1 levels are monitored for years and the incidence of complications in patients with normal levels over the course of such extended periods is compared to that in patients with persistently elevated levels, are necessary to assess the contribution of the severity and duration of hyperglycemia to the development and/or progression of complications. Nevertheless, the results of several short-term studies that have addressed facets of this issue are promising with respect to uncovering relationships between glycemic control and diabetic sequelae. For example, the significant correlation between Hb A_{1c} and cholesterol levels that has been found in most[1,88–94] but not all[95] studies would seem to indict hyperglycemia as a causative factor in macroangiopathic disease. Many investigators have also reported a relationship between serum triglycerides and Hb A_{1c} levels,[90,94] although this has not been uniformly found.[88,89] Since high-density lipoprotein (HDL) levels correlate inversely with risk for macrovascular disease, the relation between HDL levels and glucose control as assessed by measurement of glycosylated hemoglobin is also of interest. However, studies examining this relationship have yielded conflicting results.[46,89,94,96-98]

With respect to microangiopathic complications, the results of one study failed to show a correlation between muscle capillary basement membrane thickness and Hb A_{1c},[3] whereas another reported a decrease in quadriceps capillary basement membrane width after 8 to 10 months of optimum diabetic control and normal Hb A_{1c} concentrations.[99] After 2 years of treatment with continuous subcutaneous insulin infusion, glycosylated hemoglobin levels and skeletal muscle capillary basement membrane width in 13 insulin-dependent diabetic patients were both significantly reduced compared to values obtained before initiation of the intensified insulin regimen.[100] This contrasted with results in a group of 10 control patients who received conventional insulin therapy during the same interval and in whom no significant change in their glycosylated hemoglobin or capillary basement membrane width occurred. Quadriceps muscle capillary basement membrane thickness correlated with glycosylated hemoglobin, averaged from 2 to 15 determinations performed during the preceding 2½ years, in a cross-sectional analysis of 39 postpubertal youngsters with Type I diabetes.[101] Capillary basement membrane thickness and glycosylated hemoglobin also showed a positive correlation in longitudinal analysis of data from these patients. However, the same study found a negative association between glycohemoglobin and basement membrane thickness in 32 pubertal patients, and no correlation between changes in membrane thickness and glycosylated hemoglobin was found in a longitudinal analysis of this group.

Glycosylated hemoglobin was elevated in 29 diabetic patients regardless of the presence or absence of retinopathy,[102] but this was a static study in which attempt at correlation was made with only a single Hb A_1 determination. Another, similar study, performed in Pima Indians, found that the prevalence of retinopathy was greatest in those individuals in whom glycosylated hemoglobin levels were elevated, and the frequency of retinopathy increased as glycosylated hemoglobin levels increased.[45] In a group of 149 Type II diabetic patients followed from diagnosis and reassessed after 7 years, those patients with retinopathy had significantly higher random Hb A_{1c} values, and the presence of substantial retinopathy, defined as more than five microaneurysms, correlated with this glycemic index.[103] Santiago and colleagues recently reported that leakage of retinal fluorescein decreased in conjunction with normalization of Hb A_{1c} levels in a group of insulin-dependent diabetic patients.[104] Slowing of motor but not sensory nerve conduction velocity correlated with levels of glycosylated hemoglobin in 20 untreated insulin-dependent diabetic patients.[105] In six patients with insulin-dependent diabetes and an elevated glomerular filtration rate (GFR), which some view as an early marker of and/or a contributor to diabetic renal damage, the GFR fell significantly in conjunction with normalization of Hb A_1 values during a year of insulin pump therapy.[106] Increased urinary excretion of glycosaminoglycans and of an immunoelectrophoretically characterized glomerular basement membrane antigen correlated with elevated Hb A_{1c} levels in a group of 18 children with insulin-dependent diabetes.[107]

Correlative analyses of the extent of glycosylation of tissue proteins and diabetic complications in that tissue are scant. This is unfortunate because such data would help address the issue of nonenzymatic glycosylation as a contributing factor to clinical disease. The level of nonenzymatic glycosylation of collagen prepared from forearm skin biopsies from diabetic patients was significantly greater than that in control subjects.[108] However, skin collagen glycosylation was similar in diabetic patients with and without limited joint mobility, prompting the investigators to suggest that nonenzymatic glycosylation of collagen does not play an important role in the development of limited joint mobility in diabetes. In contrast, the results of another study suggested a correlation between an "index of complications" and the extent of glycosylation in tissue proteins, measured as lysine-bound glucose by furosine analysis.[109] This index was derived from evaluation of clinical records and from pathologic examination of autopsy specimens, which was the source of material for the furosine determinations, and calculated according to an arbitrary scale of 0 to 3. A history of coronary artery disease, circulatory disturbances, gangrene, or amputation was taken as an indication of macrovascular disease. The presence and severity of retinopathy was assessed by ophthalmologic exam recorded in the clinical chart. A history of paresthesias and muscular weakness or pain, and evidence of altered reflexes, were taken to reflect neuropathic disease. Glomerular disease was ascertained by histologic examination of autopsy specimens. The final index, calculated by summing the respective scores in each category, tended to rise as levels of lysine-bound glucose increased in aortic specimens (Figure 4-5).

It is clear that more such analyses are needed, with focus on specific clinical parameters and levels of glycosylation (or advanced glycosylation end products; see Chapter 5) over time in individual tissues. Unfortunately, the limited amounts and types of tissues that can be obtained by biopsy from living subjects present obstacles to these kinds of studies. A recent report describing measurement of furosine content in nail proteins offers promise for future studies designed to assess relationships between diabetic complications and tissue levels of the glycosylation product fructosyllysine[110] The investigators noted

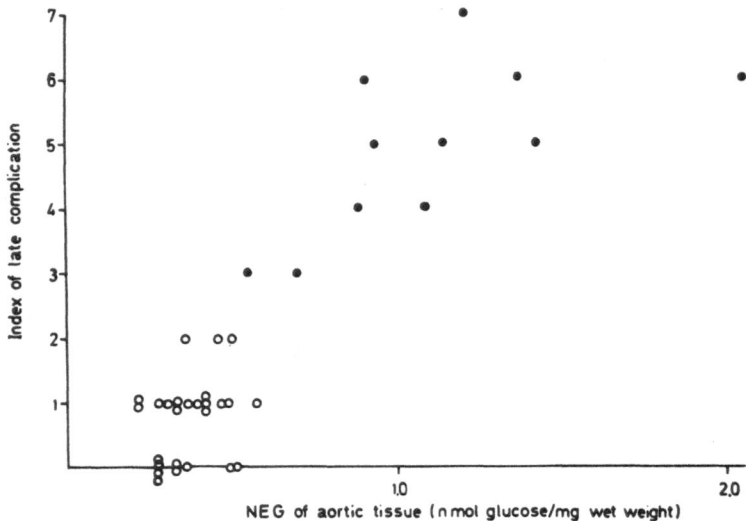

FIGURE 4-5 Complications of diabetes as a function of nonenzymatic glycosylation (NEG), measured as lysine-bound glucose by furosine analysis in autopsy specimens of aortas. O, nondiabetic subjects; ●, diabetic subjects. Reproduced with permission of the American Diabetes Association, Inc., from Vogt BW et al: ε-Amino-lysine-bound glucose in human tissues obtained at autopsy: Increase in diabetes. *Diabetes* 1982;31:1123–1127.

that nail furosine correlates with Hb A_1, that nails have a longer life span than erythrocytes, and that fingernails grow at a rate of about 4 mm/month, and they suggested that nail fructosyllysine may reflect the level of glycosylation in proteins of other tissues. Since nail samples can be collected easily and repeatedly, their analysis could provide an approach for examining the relationship between glycosylation of tissue proteins and diabetic sequelae, as well as a novel means for assessing long-term diabetic control. However, a note of caution seems appropriate since, at least according to one study,[108] glycosylation in one tissue (forearm skin) did not correlate with disease (limited joint mobility) in another, although it did correlate with Hb A_1. Measurement of glycosylated proteins, and of advanced glycosylation end products, in the affected tissue should help resolve these issues.

References

1. Gabbay KH, Hasty K, Breslow JL, et al: Glycosylated hemoglobins and long term blood glucose control in diabetes mellitus. *J Clin Endocrinol Metab* 1977;44:859–864.

2. Gonnen B, Rubenstein AH, Rochman H, et al: Hemoglobin A_1: an indicator of metabolic control of diabetic patients. *Lancet* 1977;2:734–737.

3. Koenig RJ, Peterson CM, Kilo C, et al: Hemoglobin A_{1c} as an indicator of the degree of glucose intolerance in diabetes. *Diabetes* 1976;25:230–232.

4. Koenig RJ, Peterson CM, Jones RL, et al: The correlation of glucose regulation and hemoglobin A_{1c} in diabetes mellitus. *New Engl J Med* 1976;295:417–420.

5. Gonnen B, Rochman H, Rubenstein AH: Metabolic control in diabetic patients: Assessment by hemoglobin A_1 values. *Metabolism* 1979;(suppl I):448–452.

6. Blanc MH, Barnett DM, Gleason RE, et al: Hemoglobin A_{1c} compared with three conventional measures of diabetes control. *Diabetes Care* 1981;4:349–353.

7. Lanoe R, Soria J, Thibult N, et al: Glycosylated hemoglobin concentrations and clinical test results in insulin-dependent diabetes. *Lancet* 1977;2:1156–1157.

8. Walinder O, Wibell L, Tuvemo T: Relation between hemoglobin A_1 and determinations of glucose in diabetics treated with and without insulin. *Diabete Metab* 1980;6:251–255.

9. Goldstein DE, Walker B, Rawlings SS, et al: Hemoglobin A_{1c} levels in children and adolescents with diabetes mellitus. *Diabetes Care* 1980;3:503–507.

10. Agardh C-D, Tallroth G: Lack of correlation between glycosylated haemoglobin concentrations and number of daily insulin injections: cross sectional study in care of ambulatory diabetes. *Br Med J* 1985; 291:622.

11. Mecklenberg RS, Benson EA, Benson JW Jr, et al: Long-term metabolic control with insulin pump therapy. Report of experience with 127 patients. *N Engl J Med* 1985;313:465–468.

12. Dahlquist G, Blom L, Bolme P, et al: Metabolic control in 131 juvenile-onset diabetic patients as measured by HbA_{1c}: Relation to age, duration, C-peptide, insulin dose, and one or two insulin injections. *Diabetes Care* 1982;5:399–403.

13. Mazze RS, Shamoon H, Pasmantier R, et al: Reliability of blood glucose monitoring by patients with diabetes mellitus. *Am J Med* 1984;77:211–217.

14. Citrin W, Ellis GJ, Skyler JS: Glycosylated hemoglobin: A tool in identifying psychological problems. *Diabetes Care* 1980;3:563–564.

15. McDermott K, Cooks M, Peterson CM: Patient determined glycosylated hemoglobin measurements: An aid to patient education.*Diabetes Care* 1981;4:480–483.

16. Dunn, PJ, Cole RA, Soeldner JS, et al: Stability of hemoglobin A_{1c} levels on repetitive determination in diabetic outpatients. *J Clin Endo Metab* 1981;52:1019–1022.

17. Schultz TA, Lewis SB, Davis JL, et al: Effect of sulfonylurea therapy and plasma glucose levels on hemoglobin A_{1c} in Type II diabetes mellitus. *Am J Med* 1981;70:373–378.

18. Turner RC, et al (a multicenter study): U.K. Prospective Diabetes Study:

II. Reduction in HbA_{1c} with basal insulin supplement, sulfonylurea, or biguanide therapy in maturity-onset diabetes. *Diabetes* 1985;34:793-798.

19. Pecoraro RE, Chen MS, Porte D: Glycosylated hemoglobin and fasting plasma glucose in the assessment of outpatient glycemic control in NIDDM. *Diabetes Care* 1982;5:592-599.

20. Graf RJ, Halter JB, Porte D: Glycosylated hemoglobin in normal subjects and subjects with maturity onset diabetes. *Diabetes* 1978;27:834-839.

21. Dolhofer R, Wieland OH: Glycosylation of serum albumin: Elevated glucosyl albumin in diabetic patients. *FEBS Lett* 1979;103:282-286.

22. Dolhofer R, Wieland OH: Increased glycosylation of serum albumin in diabetes mellitus. *Diabetes* 1980;29:417-422.

23. Guthrow CE, Morris MA, Day JF, et al: Enhanced nonenzymatic glucosylation of serum albumin in diabetes mellitus. *Proc Natl Acad Sci USA* 1979;76:4258-4261.

24. Day JF, Thornburg RW, Thorpe SW, et al: Nonenzymatic glucosylation of rat albumin: Studies in vitro and in vivo. *J Biol Chem* 1979;254:9394-9400.

25. Day JF, Ingelbretsen CG, Ingelbretsen WR, et al: Nonenzymatic glucosylation of serum proteins and hemoglobin: Response to changes in blood glucose levels in diabetic rats. *Diabetes* 1980;29:524-527.

26. Day JF, Thorpe SR, Baynes JW: Non-enzymatically glucosylated albumin: In vitro preparation and isolation from normal human serum. *J Biol Chem* 1979;254:595-597.

27. Dolhofer R, Renner R, Wieland OH: Different behavior of haemoglobin A_{1a-c} and glucosyl-albumin levels during recovery from diabetic ketoacidosis and non-acidotic coma. *Diabetologia* 1981:21;211-215.

28. Kemp SF, Creech RH, Horn TR: Glycosylated albumin and transferrin: Short term markers of blood glucose control. *J Pediatr* 1984;105:394-398.

29. McFarland KF, Catalano EW, Day JF, et al: Nonenzymatic glucosylation of serum proteins in diabetes mellitus. *Diabetes* 1979;28:1011-1014.

30. Yue DK, Morris K, McLennan S, et al: Glycosylation of plasma protein and its relation to glycosylated hemoglobin in diabetes. *Diabetes* 1980;29:296-300.

31. Kennedy Al, Kandell TW, Merimee TJ: Serum protein-bound hexose in diabetes: The effect of glycemic control. *Diabetes* 1979;28:1006-1010.

32. Gragnoli G, Tanganelli I, Signorini AM, et al: Nonenzymatic glycosylation of serum protein as an indicator of diabetic control. *Acta Diabetol Lat* 1982;19:161-166.

33. Jones IR, Owens DR, Williams S, et al: Glycosylated serum albumin: An intermediate index of diabetic control. *Diabetes Care* 1983;6:501-503.

34. Kennedy AL, Mehl TD, Merimee TJ: Nonenzymatically glycosylated serum protein: Spurious elevation due to free glucose in serum. *Diabetes* 1980;29:413-415.

35. Rendell M, Kao G, Mecherikunnel P, et al: Use of aminophenylboronic acid affinity chromatography to measure glycosylated albumin levels. *J Lab Clin Med* 1985;105:63-69.

36. Willey DG, Rosenthal MA, Caldwell S: Glycosylated haemoglobin and plasma glycoprotein assays by affinity chromatography. *Diabetologia* 1984;27:56–58.

37. Mehl TD, Wenzel SE, Russell B, et al: Comparison of two indices of glycemic control in diabetic subjects: Glycosylated serum protein and hemoglobin. *Diabetes Care* 1983;6:34–39.

38. Tsuchiya S, Sakurai T, Sekiguchi S-I: Nonenzymatic glucosylation of human serum albumin and its influence on binding capacity of sulfonylureas. *Biochem Pharmacol* 1984;33:2967–2971.

39. Shaklai N, Garlick RL, Bunn HF: Nonenzymatic glycosylation of human serum albumin alters its conformation and function. *J Biol Chem* 1984;259:3812–3817.

40. Brownlee M, Vlassara H, Cerami A: Measurement of glycosylated amino acids and peptides from urine of diabetic patients using affinity chromatoghraphy. *Diabetes* 1980;29:1044–1047.

41. Gragnoli G, Signorini AM, Tanganelli I: Nonenzymatic glycosylation of urinary proteins in Type I (insulin-dependent) diabetes: Correlation with metabolic control and degree of proteinuria. *Diabetologia* 1984;26:411–414.

42. Lev-Ran A: glycohemoglobin : Its use in the follow-up of diabetes and diagnosis of glucose intolerance. *Arch Intern Med* 1981;141:747–749.

43. Santiago JV, Davis JE, Fisher F: Hemoglobin A_{1c} levels in a diabetes detection program. *J Clin Endocrinol Metab* 1978;47:578–580.

44. Dunn PJ, Cole RA, Soeldner JS, et al: Temporal relationship of glycosylated hemoglobin concentrations to glucose control in diabetes. *Diabetologia* 1979;17:213–220.

45. Flock EV, Bennett PH, Savage PJ, et al: Bimodality of glycosylated hemoglobin distribution in Pima Indians. *Diabetes* 1979;28:984–989.

46. Dods RF, Bolmey C: Glycosylated hemoglobin assay and oral glucose tolerance test compared for detection of diabetes mellitus. *Clin Chem* 1979;25:764–768.

47. Simon D, Coignet MC, Thibult N, et al: Comparison of glycosylated hemoglobin and fasting plasma glucose in the detection of diabetes mellitus. *Am J Epidemiol* 1985;122:589–593.

48. Dunn PJ, Cole RA, Soeldner JS, et al: Reproducibility of hemoglobin A_{1c} and sensitivity to various degrees of glucose intolerance. *Ann Intern Med* 1979;91:390–396.

49. Lev-Ran A, Vanderlaan WP: Glycohemoglobins and glucose intolerance. *JAMA* 1979;241:912–914.

50. Boucher BJ, Welch SG, Beer MS: Glycosylated haemoglobins in the diagnosis of diabetes mellitus and for the assessment of chronic hyperglycemia. *Diabetologia* 1981;21:34–36.

51. Kesson CM, Young RE, Talwar D, et al: Glycosylated hemoglobin in the diagnosis of non-insulin dependent diabetes mellitus. *Diabetes Care* 1982;5:395–398.

52. Bolli G, Compagnucci P, Cartechini MG, et al: HbA_1 in subjects with abnormal glucose tolerance but normal fasting plasma glucose. *Diabetes* 1980;29:272–277.

53. Dix D, Cohen P, Kingsley S, et al: Glycohemoglobin and glucose tolerance tests compared as indicators of borderline diabetes. *Clin Chem* 1979;25:877–879.
54. Cederholm J, Ronquist G, Wibell L: Comparison of glycosylated hemoglobin with oral glucose tolerance test. *Diabete Metab* 1984;10:224–229
55. Hall PM, Cook JGH, Sheldon J, et al: Glycosylated hemoglobins and glycosylated plasma proteins in the diagnosis of diabetes mellitus and impaired glucose tolerance. *Diabetes Care* 1984;7:147–150.
56. Walinder O, Libell L, Tuvemo T: Relation between hemoglobin A$_{1c}$ and determintions of glucose in diabetics treated with and without insulin. *Diabete Metabol* 1980;6:251–255.
57. Gomo ZAR: The determination of glucose and glycosylated haemoglobin in a nondiabetic Zimbabwean African population. *Ann Clin Biochem* 1985;22:362–365.
58. Soler NG, Frank S: Value of glycosylated hemoglobin measurements after myocardial infarction. *JAMA* 1981;246:1690–1693.
59. Schleicher ED, Gerbitz KD, Dolhofer R, et al: Clinical utility of nonenzymatically glycosylated blood proteins as an index of glucose control. *Diabetes Care* 1984;7:548–558.
60. Widness JA, Schwartz HC, Kahn CB, et al: Glycohemoglobin in diabetic pregnancy: A sequential study. *Am J Obstet Gynecol* 1980;136:1024–1029.
61. Lind T, Cheyne GA: Effect of normal pregnancy upon the glycosylated hemoglobins. *Br J Obstet Gynecol* 1979;86:210–213.
62. Kjaergaard J-J, Ditzel J: Hemoglobin A$_{1c}$ as an index of long-term blood glucose regulation in diabetic pregnancy. *Diabetes* 1979;28:694–696.
63. Gillmer MDG, Beard RW, Brooks FW, et al: Carbohydrate metabolism in pregnancy. I Diurnal plasma glucose profile in normal and diabetic women. *Br Med J* 1975;3:399–404.
64. Schwartz HC, King KC, Schwartz AL, et al: Effects of pregnancy on hemoglobin A$_{1c}$ in normal, gestational diabetic, and diabetic women. *Diabetes* 1976;25:1118–1122.
65. Leslie RDG, Pyke Da, John PN, et al: Haemoglobin AI in diabetic pregnancy. *Lancet* 1978;2:958–959.
66. Miller JM Jr, Crenshaw C Jr, Welt SI: Hemoglobin A$_{1c}$ in normal and diabetic pregnancy. *JAMA* 1979;242:2785–2787.
67. Paulsen EP, Khoury M: Hemoglobin A$_{1c}$ levels in insulin-dependent diabetes mellitus. *Diabetes* 1976;25:890–896.
68. Pedersen J, Bojsen-Moller B, Poulsen H: Blood sugar in newborn infants of diabetic mothers. *Acta Endocrinol* 1954;15:33–36.
69. Freinkel N: Of pregnancy and progeny. *Diabetes* 1980;29:1023–1035.
70. Ogata ES, Freinkel N, Metzger BE, et al: Perinatal islet function in gestational diabetes: Assessment by cord plasma, C-peptide and amniotic fluid insulin. *Diabetes Care* 1980;3:425–429.
71. Susa JB, McCormick KL, Widness JA, et al: Chronic heperinsulinemia in the fetal rhesus monkey: Effects on fetal growth and composition. *Diabetes* 1979;28:1058–1063.

72. Susa JB, Neave C, Sehgal P, et al: Chronic hyperinsulinemia in the fetal rhesus monkey: Effect of physiologic hyperinsulinemia on fetal growth and composition. *Diabetes* 1984;33:656–660.
73. Hill DE: Fetal effects of insulin. *Obstet Gynecol Ann* 1982;11:133–149.
74. Susa JB, Widness JA, Hintz R, et al: Somatomedins and insulin in diabetic pregnancies: Effects on fetal macrosomia in the human and rhesus monkey. *J Clin Endocrinol Metab* 1984;58:1099–1105
75. Susa JB, Schwartz R: Effects of hyperinsulinemia in the primate fetus. *Diabetes* 1985;34(suppl 2):36–41.
76. Widness JA, Schwartz HC, Thompson D, et al: Glycohemoglobin (Hb A_{1c}): a predictor of birth weights in infants of diabetic mothers. *J Pediatr* 1978;92:8–12.
77. Baxi L, Barad D, Reece EA, et al: Use of glycosylated hemoglobin as a screen for macrosomia in gestational diabetes. *Obstet Gynecol* 1984; 64:347–350.
78. O'Shaughnessy R, Russ, J, Zuspan FP: Glycosylated hemoglobins and diabetes mellitus in pregnancy. *Am J Obstet Gynecol* 1979;135:783–790.
79. Sosenko JM, Kitzmiller JL, Fluckiger R, et al: Umbilical cord glycosylated hemoglobin in infants of diabetic mothers: Relationships to neonatal hypoglycemia, macrosimia, and cord serum C-peptide. *Diabetes Care* 1982;5:566–577.
80. Madsen H, Ditzel J, Hansen P, et al: Hemoglobin A_{1c} Determinations in diabetic pregnancy. *Diabetes Care* 1981;4:541–546.
81. Jovanovic L, Petersen CM: The clinical utility of glycosylated hemoglobin. *Am J Med* 1981;70:331–338.
82. Pedersen JF, Molsted-Pedersen L, Mortensen HB: Fetal growth delay and maternal hemoglobin A_{1c} in early diabetic pregnancy. *Obstet Gynecol* 1984;64:351–352.
83. Miller E, Hare JW, Cloherty JP, et al: Elevated maternal hemoglobin A_{1c} in early pregnancy and major congenital anomalies in infants of diabetic mothers. *N Engl J Med* 1981;304:1331–1334.
84. Ylinen K, Aula P, Stenman UH, et al: Risk of minor and major fetal malformations in diabetics with high haemoglobin A_{1c} values in early pregnancy. *Br Med J* 1984;289:345–346.
85. Sosenko IR, Kitzmiller JL, Loo SW, et al: The infant of the diabetic mother. Correlation of increased cord C-peptide levels with macrosomia and hypoglycemia. *N Engl J Med* 1979;301:859–862.
86. McFarland KF, Murtiashaw M, Baynes JW: Clinical values of glycosylated serum protein and glycosylated hemoglobin levels in the diagnosis of gestational diabetes mellitus. *Obstet Gynecol* 1984;64:516–518.
87. Fadel HE, Hammond SD, Huff TA, et al: Glycosylated hemoglobins in normal pregnancy and gestational diabetes mellitus. *Obstet Gynecol* 1979;54:322–326.
88. Chase HP, Glasgow AM: Juvenile diabetes mellitus and serum lipids and lipoprotein levels. *Am J Dis Child* 1976;130:1113–1117.
89. Klubjer L, Malnar D, Kardos M, et al: Metabolic control, glycosylated

hemoglobin, and high density lipoprotein cholesterol in diabetic children. *Eur J Pediatr* 1979;132:289-297.

90. Moore WV, Knapps J, Kauffmann Rl, et al: Plasma lipid levels in insulin-dependent diabetes mellitus. *Diabetes Care* 1979;2:31-34.

91. Nikkila EA, Hormila P: Serum lipids and lipoproteins in insulin-treated diabetes. *Diabetes* 1976;27:1078-1086.

92. Peterson CM, Koenig RJ, Jones RL, et al: Correlation of serum triglyceride levels and HbA_{1c} concentrations in diabetes mellitus. *Diabetes* 1977;26:507-509.

93. Sosenko JM, Breslow JL, Muttinew OS, et al: Hyperglycemia and plasma lipid levels: A prospective study of young insulin-dependent diabetic patients. *N Engl J Med* 1980;302:650-654.

94. Aleyassine H, Gardiner RJ, Tonks DB, et al: Glycosylated hemoglobin in diabetes mellitus: Correlations with fasting plasma glucose, serum lipids and glycosuria. *Diabetes Care* 1980;3:508-514.

95. Elkeles RJ, Wu J, Hambley J: Hemoglobin A_1 blood glucose and high density lipoprotein cholesterol in insulin requiring diabetes. *Lancet* 1978;2:547-548.

96. Calvert GD, Graham JJ, Mannick T, et al: Effects of therapy on plasma high-density lipoprotein-cholesterol concentration in diabetes mellitus. *Lancet* 1978;2:66-68.

97. Kennedy Al, Lappin TRJ, Lavery TD, et al: Relation of high density lipoprotein cholesterol concentration to type of diabetes and its control. *Br Med J* 1978;2:1191-1194.

98. Lopes-Virella MFL, Stone PG, Colwell JA: Serum high density lipoprotein in diabetic patients. *Diabetologia* 1977;13:285-291.

99. Petersen CM, Jones Rl, Esterly JA, et al: Changes in basement membrane thickening and pulse volumes concomitant with improved glucose control and exercise in patients with insulin-dependent diabetes mellitus. *Diabetes Care* 1980;3:586-589.

100. Raskin P, Pietri AO, Unger R, et al: The effect of diabetic control on the width of skeletal-muscle capillary basement membrane in patients with Type I diabetes mellitus. *N Engl J Med* 1983;309:1546-1550.

101. Sosenko JM, Miettinen OS, Williamson JR, Gabbay KH: Muscle capillary basement membrane thickness and long-term glycemia in Type I diabetes mellitus. *N Engl J Med* 1984;311:694-698.

102. Coller BS, Frank RN, Milton RC, Gralnick HR: Plasma cofactors of platelet function: Correlation with diabetic retinopathy and hemoglobin A_{1a-c}. *Ann Intern Med* 1978;88:311-316.

103. Williams JH, Hillson RM, Bron A, et al: Retinopathy is associated with higher glycemia in maturity-onset diabetes. *Diabetologia* 1984;27:198-202.

104. White NJ, Waltman SR, Krupin T, et al: Reversal of early ocular abnormalities in juvenile diabetics (IDD) after normalization of hemoglobin A_{1c}. *Clin Res* 1981;462A.

105. Graf RJ, Halter JB, Halar E, et al: Nerve conduction abnormalities in

untreated maturity-onset diabetes: Relation to levels of fasting plasma glucose and glycosylated hemoglobin. *Ann Intern Med* 1979;90:298–303.

106. Wiseman MJ, Saunders AJ, Keen H, et al: Effect of blood glucose control on increased glomerular filtration rate and kidney size in insulin-dependent diabetes. *N Engl J Med* 1985;312:617–621.

107. Lubec G, Legenstein E, Pollack A, et al: Glomerular basement membrane changes, HbA$_{1c}$ and urinary excretion of acid glycosaminoglycans in children with diabetes mellitus. *Clin Chim Acta* 1980;103:45–49.

108. Lyons TJ, Kennedy L: nonenzymatic glycosylation of skin collagen in patients with Type I (insulin-dependent) diabetes mellitus and limited joint mobility. *Diabetologia* 1985;28:2–5.

109. Vogt BW, Schleicher ED, Wieland OH: ε-amino-lysine-bound glucose in human tissues obtained at autopsy: Increase in diabetes mellitus. *Diabetes* 1982;31:1123–1127.

110. Oimomi M, Hatanaka H, Ishikawa K, et al: Increased fructose-lysine of nail protein in diabetic patients. *Klin Wochenschr* 1984;62:477–478.

CHAPTER 5

Pathophysiologic Significance

Blood Components

Hemoglobin

The amino-terminal of the β chain of hemoglobin, to which glucose attaches, is also a site where organic phosphates bind. 2,3-Diphosphoglycerate (2,3-DPG), an important intermediate in red cell glycolysis, influences the affinity of hemoglobin for oxygen through its ability to bind to β chain residues of deoxyhemoglobin. Addition of organic phosphate decreases the oxygen affinity of hemoglobin, whereas removal of organic phosphate increases the oxygen affinities of Hb A and Hb A_{1c}. The availability of this site for interaction with 2,3-DPG is compromised when it is covalently linked to glucose.[1] Hence, Hb A_{1c} exhibits greater oxygen affinity than Hb A in the presence of 2,3-DPG.[2-4] Correspondent with the increase in the level of Hb A_{1c} in the red cells of diabetic patients, the oxygen affinity of these cells in the presence of 2,3-DPG is slightly greater than that of red cells from nondiabetic subjects.[5] This difference could be explained on the basis of the interference presented by the NH_2-terminal glucose of Hb A_{1c} to the binding of 2,3-DPG.

The impact of this minor shift in the hemoglobin-oxygen dissociation curve arising from increased Hb A_{1c} in the red cells of diabetic persons is not clear. Ditzel suggested that it promotes tissue hypoxia, which in turn could contribute to diabetic complications,[6-9] although this interpretation rests on the extrapolation of in vitro data to in vivo situations. One study found an increased hemoglobin-oxygen affinity in blood from diabetic children,[10] and it is possible that increased Hb A_{1c} coupled with low red cell 2,3-DPG in diabetes,[11,12] compromises oxygen delivery to tissues. However, such postulates must be reconciled with the finding that the modest increase in hemoglobin oxygen affinity following meals in diabetic children correlates with increments in blood pH but not with the rise in Hb A_{1c},[13] and the report that the slightly higher pO_2 at which hemoglobin is half-saturated with oxygen in diabetic samples is independent of the state of glycosylation of hemoglobin and is probably of little physiologic significance.[14] Furthermore, as Bunn and colleagues have pointed out,[15,16] the fact that certain hemoglobinopathies produce greater shifts in the oxygen dissociation curve without deleterious effects on tissue oxygenation makes it unlikely that the minor change in oxygen affinity associated with increased Hb A_{1c} in diabetes significantly alters oxygen transport. Thus, although investigations of glycosylated hemoglobins have provided critical insights into structure-function relationships, the principal clinical contribution derived from studies of the minor hemoglobin components in diabetes remains the conclusive demonstration that their measurement can reliably assess integrated blood glucose levels over the preceding 6 weeks.

Albumin

Like Hb A_1, measurement of the level of glycosylation of serum albumin has undisputed clinical value because of its ability to provide an index of blood glucose control during the preceding period of 1 to 4 weeks. However, there is now also evidence implicating the non-enzymatic glycosylation of albumin in the pathophysiology of diabetic microangiopathy be mechanisms quite distinct from those invoked for glycosylated hemoglobin. Although nonenzymatic glycosylation of albumin does not affect its circulating half-life or catabolism,[17] it can induce a conformational change in the protein and alter its ligand-binding properties.[18] The glycosylated form appears to be taken up more avidly than native albumin by endothelial cells[19] This process involves micropinocytic vesicles of endothelial cells, which participate in the bidirectional transport of proteins across the capillary wall. Transendothelial vesicular ingestion of nonenzymatically glucosylated myoglobin and ovalbumin is also enhanced compared to the non-modified forms of these proteins[20](Figure 5-1), and glucosylated

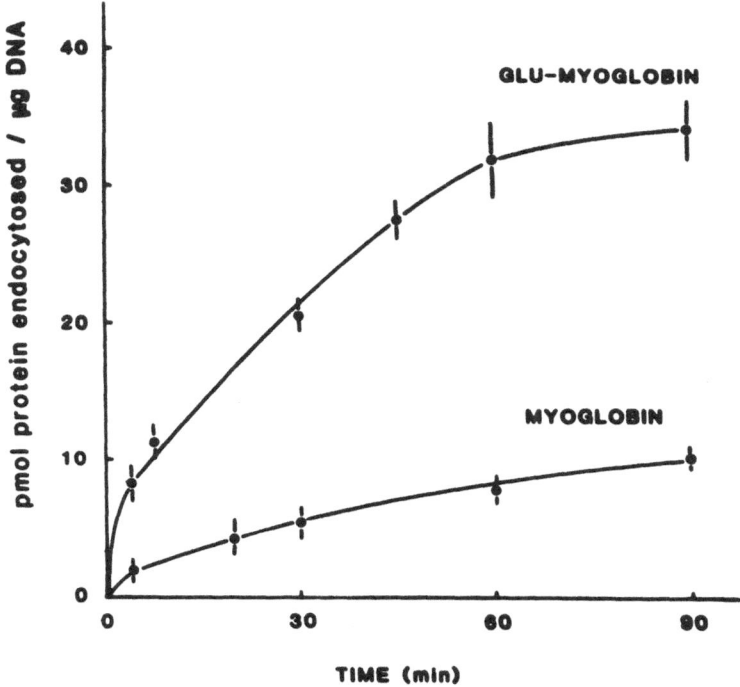

FIGURE 5-1 Enhancement of vesicular ingestion of myoglobin by capillary endothelial cells as a result of nonenzymatic glucosylation. Reprinted with permission from Williams SK, Solenski NJ: Enhanced vesicular ingestion of nonenzymatically glycosylated proteins by capillary endothelium. *Microvasc Res* 1984;28:311–321. Academic Press, New York.

ferritin appears to be preferentially tranported across the glomerular basement membrane filtration barrier.[21] In the latter experiments, which entailed electron microscopy of the glomerular capillary wall after perfusion of ferritin into rat kidneys, carbohydrate-free ferritin accumulated within the lamina rara interna and was restricted from transglomerular transport, whereas glucosylated ferritin penetrated the lamina densa and the lamina rara externa and accumulated in epithelial pinocytic vesicles and multivesicular bodies. Thus, glycosylation of albumin and other circulating proteins in uncontrolled diabetes could increase transendothelial transport and contribute to the increased capillary permeability associated with diabetes. In support of this concept is the finding that the passage of glycosylated albumin through the glomerular filtration barrier is enhanced relative to that of normal albumin in diabetic patients without and with microalbuminuria as well as in normal people.[22] In diabetic subjects, the clearance of albumin correlated with the serum concentration of glycoalbumin, and the urinary-to-serum glycoalbumin ratio correlated inversely with the albumin clearance. Of additional interest is the report that repeated intravenous injection of glucosylated plasma proteins produces glomerular basement membrane thickening in nondiabetic mice.[23] This does not appear to be due to binding of exogenously administered glycosylated albumin to the basement membrane,[24] although binding of endogenous albumin is known to occur.[25] However, when actual measurements of glomerular basement membrane width were performed by another group of investigators, no effect of repeated injections of glucosylated plasma proteins was observed.[26]

Glycosylated albumin containing 35 mole of glucose per mole of albumin was able to inhibit the uptake by rat hepatic endothelial cells of ^{125}I-labeled agalactoorosomucoid (AGOR), a glycoprotein carrying N-acetylglucosamine as the terminal carbohydrate residue.[27] Free glucose also inhibited the uptake of AGOR by endothelial cells, but the glucosyl albumin conjugate was about 6 orders of magnitude more potent an inhibitor. The significance of this finding resides in the fact that the terminal nonreducing sugar residue of the oligosaccharide moiety of glycoproteins provides a carbohydrate recognition factor for their receptor-mediated uptake by hepatic cells.[28] This in turn influences the survival of glycoproteins such as orosomucoid in the circulation and their clearance by the liver.[29,30] Thus, nonenzymatic glycosylation of albumin or other circulating proteins could compromise hepatic uptake of those glycoproteins that depend on carbohydrate recognition for their clearance.

Globulins

Incubation of normal human serum with glucose or galactose results in the nonenzymatic glycosylation of IgG and IgM as well as albumin.[31,32] This is not surprising in view of the fact that in vitro nonenzymatic glycosylation has been demonstrated with virtually every protein that has been examined to date. However, the potential significance proposed by the investigators reporting this finding is of interest. They note that sepsis neonatorum is common in galactosemic infants, and therefore suggest that immunologic function may be impaired with galactosemia or with hyperglycemia as a result of changes in the biologic activity of IgG imposed by nonenzymatic glucosylation. On the other hand, in vitro glycosylation of α_2-macroglobulin to levels of 10.3 mol of glucose per mole of protein did not alter its binding to monolayer cultures of mouse peritoneal macrophages.[33] The clearances of nonglycosylated and glycosylated forms of methylamine-treated α_2-macroglobulin, assessed by disappearance curves after tail vein injection into mice, were also identical. These findings were surprising in view of the fact that lysine residues of α_2-macroglobulin participate in recognition and binding of the protein to its cell surface receptor,[34] although it is possible that the critical lysine that is involved in receptor recognition did not become glycosylated in these experiments.

Lipoproteins

Nonenzymatic glycosylation of low-density lipoproteins (LDL) has been pathogenetically implicated in the increased incidence of macrovascular disease associated with diabetes. Incubation of human serum lipoproteins with glucose or galactose in vitro results in the covalent binding of glucose to ε-amino groups of lysine residues in the apolipoproteins of LDL, VLDL (very-low-density lipoproteins), and HDL (high-density lipoproteins); glycosylation of the apolipoprotein B of LDL purified from serum of diabetic patients is increased compared to levels of LDL glycosylation in samples from nondiabetic subjects.[35-38] Binding and degradation of glycosylated LDL by cultured human fibroblasts and by umbilical vein endothelial cells is diminished compared to nonglycosylated LDL (Figure 5-2), and the

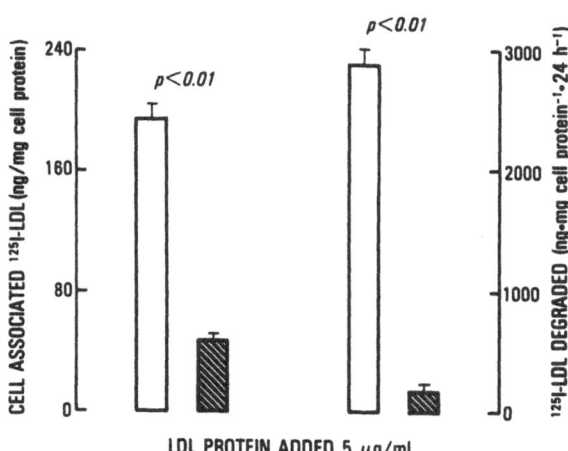

FIGURE 5-2   Inhibition of uptake and degradation of LDL by endothelial cells as a result of nonenzymatic glucosylation. Open columns, control LDL; hatched columns, glucosylated LDL (16% of lysyl residues modified). Reprinted with permission from Lorenzi et al.[41]

degree of reduction in degradation is greater with increasing extent of glycosylation.[38-41] The glycosylation of LDL abolishes the high-affinity uptake and degradation process by normal skin fibroblasts, and also results in a decreased rate of clearance in vivo,[38,42,43] (Figure 5-3). In contrast to native LDL, which inhibits the activity of β-hydroxy-methylglutaryl coenzyme A reductase (HMG-CoA reductase) and stimulates acyl-CoA:cholesterol acyltransferase, glycosylated LDL does not affect these enzymes.[38] Thus, disturbances in receptor-mediated internalization and degradation of LDL as a result of glycosylation also interferes with intracellular handling of cholesterol and regulation of its synthesis. Such findings prompted the suggestion that glycosylation of lysyl residues on apo-LDL, which is intimately involved with LDL receptor recognition, as a consequence of persistent hyperglycemia could alter LDL metabolism and contribute to atherogenesis in diabetic subjects.

One recent report casts some doubt on the pathobiologic role of these findings. Arguing that many of the in vitro studies used

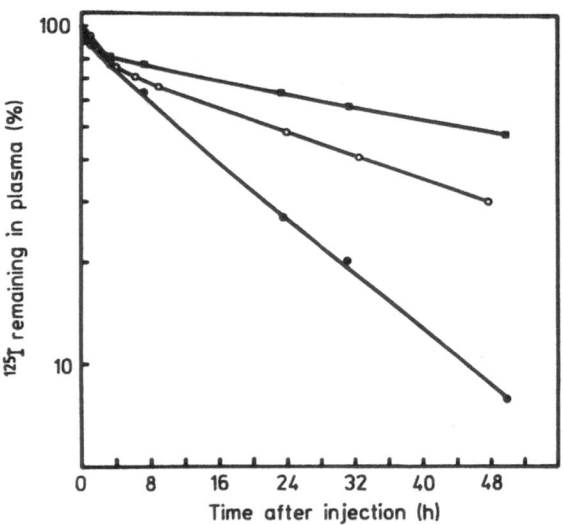

FIGURE 5-3 Clearance of ^{125}I-labeled LDL and glycosylated ^{125}I-labeled LDL from rabbit plasma. ●, native LDL; O, glucosylated LDL (23% of lysyl residues modified); ■, galactosylated LDL (69% of lysyl residues modified). Reprinted with permission from Sasaki J et al: Measurement of receptor-independent metabolism of low-density lipoprotein: An application of glycosylated low-density lipoprotein. *Eur J Biochem* 1983;131:535–538.

glucosylated LDL that had been reduced with sodium borohydride and hence represented unphysiologic conditions, Schleicher et al. examined the effect of nonreductive nonenzymatic glucosylation on the uptake and degradation of LDL in fibroblasts, hepatocyte-macrophages, and endothelial cells.[44] With these conditions, moderate glucosylation of LDL (1.8 to 4.6 glucosylated lysine residues per molecule of apolipoprotein B) did not alter its interaction with the high-affinity receptor present on various cell types. Only more heavily glucosylated LDL (>12 glucosylated residues per molecule of apolipoprotein B) or glucosylated LDL that had been subjected to reduction with sodium borohydride exhibited reduced receptor-mediated internalization and degradation by human fibroblasts. Nonreductive glycosylation at a level of 3.3 mol of glucose per mole of apolipoprotein B did not increase cholesterol content or [^{14}C]oleate incorporation in human monocyte-derived macrophages.

The above postulates also have to be reconciled with the report that LDL obtained from hyperglycemic patients with non-insulin-dependent diabetes reportedly binds to fibroblasts with binding affinities and binding capacities similar to those observed with LDL obtained from nondiabetic individuals.[45] On the other hand, the internalization and degradation of LDL from patients with insulin-dependent diabetes and poor metabolic control is decreased compared to that of LDL isolated from normal subjects or from insulin-dependent diabetic patients with good metabolic control.[46] Further, in vitro modification of only 2% to 5% of lysine residues in LDL, a modest level of glycosylation comparable to that occurring in some diabetics, is sufficient to produce demonstrable inhibition of LDL degradation by cultured fibroblasts and of the turnover (fractional catabolic rate) of LDL injected into guinea pigs.[47]

There is little doubt that excess nonenzymatic glycosylation of LDL occurs in vivo in diabetic patients with poor metabolic control. In an elegant study that utilized a panel of mouse monoclonal antibodies generated against glucosylated LDL and employed competitive double-antibody radioimmunoassays, Curtiss and Witztum demonstrated that the plasma apolipoproteins A1, A11, B, C1, and E all become glycosylated in hyperglycemic diabetic subjects.[48,49] The overall level of nonenzymatic glycosylation of lipoproteins in the plasma of poorly controlled diabetic patients was up to 33 times greater than that in plasma lipoproteins of nondiabetic individuals. In diabetic samples, triglyceride-rich lipoproteins contained the greatest number of glucitolysine residues, whereas most of the glycosylated lipoprotein in normal plasma was recovered in the HDL fractions. Coupled with results of in vitro studies, which documented that transfer of glucosylated apoproteins from HDL to VLDL can occur, these findings suggested to the investigators that the glycosylated apolipoproteins recovered in the VLDL fraction of plasma from hyperglycemic patients had been transferred from the HDL class.

There are several ways in which the above changes in LDL metabolism due to nonenzymatic glycosylation might promote atherogenesis. For example, decreased clearance could cause elevated plasma LDL levels. Further, glycosylated LDL might be taken up at tissue sites other than those normally involved with LDL degradation, or it might be metabolized differently than native LDL. Recent studies indicating that LDL catabolism occurs through receptor-independent mechanisms as well as receptor-dependent pathways provide evidence for the latter possibility. Modification of arginine or lysine residues of LDL by cyclohexanedione or reductive methylation will, like glycosylation, block uptake of LDL by the LDL receptor. These modified lipoproteins are metabolized through receptor-independent processes.[50,51] Receptor-independent catabolism occurs in a variety of tissues, notably in cells of the reticuloendothelial system such as monocyte-macrophages. Additionally, macrophages can take up LDL by another receptor-mediated pathway that specifically recognizes chemically modified LDL. Although the original description of macrophage receptors for modified lipoproteins involved acetylated LDL,[52] this process may be operative for naturally modified lipoproteins such as would result from nonenzymatic glycosylation in vivo. Since these receptors are not regulated by cholesterol requirements of the cell, macrophages can accumulate large quantities of LDL by this route. If employment of these receptors to remove modified LDL from plasma or tissue occurs in vivo as part of the scavenger function of macrophages, glycosylation of LDL in diabetic subjects could have atherogenic potential in view of fact that arterial wall macrophages give rise to the foam cells of atherosclerotic plaques. Again, this postulate must be reconciled with reports that LDL obtained from patients with non-insulin-dependent diabetes, like that from control subjects, is not bound or degraded by mouse peritoneal macrophages,[45] that native and glycosylated LDL are taken up and degraded at similar rates by a low-affinity process in peritoneal macrophages isolated from mice,[43] and that nonreduced glucosylated LDL does not promote accumulation of cholesteryl esters in human monocyte-derived macrophages.[44]

Coagulation Proteins

Several pathophysiologic consequences of excess nonenzymatic glycosylation of factors involved in blood clotting have been suggested. Some are quite speculative. For example, since the carbohydrate portion of Von Willebrand's factor is a major determinant of its interaction with platelets,[53] and since that of Factor V is essential for its coagulation activity,[54] it has been proposed that increased glycosylation could enhance the activity of these coagulation proteins. For other proposed conquences, some suggestive or supporting experimental evidence exists. For example, the amount of glucose bound to lysine residues in fibrinogen purified from the plasma of poorly controlled insulin-dependent diabetics is about 1½ times that bound to fibrinogen purified from nondiabetic subjects.[55] Since lysine is the amino donor for fibrin cross-linking, and since fibrin and fibrinogen (from which the fibrinopeptides A and B are split by the action of thrombin to form fibrin) are glycosylated to a similar extent, it is possible that glycosylation of lysine residues interferes with fibrinogen and/or fibrin function or processing. One such suggested effect of the attachment of glucose would be to compromise the availability of lysine residues in fibrin to the fibrin-stabilizing factor (Factor XIII) and, hence, the formation of ε-(γ-glutamyl)-lysine intramolecular cross-links. However, a recent study examining the effects of in vitro glycosylation of fibrinogen did not confirm this postulate, in that glycosylation at a level of 3.8 mol of glucose per mole of protein altered neither clotting time nor Factor XIII cross-linking of fibrinogen.[33] Glycosylation of fibrin achieved by in vitro incubation with glucose results in a reduced susceptibility to degradation by the fibrinolytic enzyme plasmin.[56] Since lysyl residues of fibrin are specific sites of plasmin hydrolysis, excess nonenzymatic glycosylation in diabetes could interfere with the degradation of deposited fibrin and lead to its accumulation in tissues. A high level of glycosylation is apparently required to produce this effect on plasmin-mediated fibrinolysis, since in another study the susceptibility to plasmin degradation of fibrin formed from fibrinogen glycosylated at a level of 3.8 mol/mol was not reduced compared to that formed from native fibrinogen.[33] However, even this level of glycosylation was sufficient to slow the rate of cleavage of fibrinogen by plasmin. Figure 5-4 depicts certain pathophysiologic consequences that oculd ensue if any of the proposed functional alterations of coagulation proteins induced by nonenzymatic glycosylation occur in vivo.

Antithrombin III is a coagulation-regulatory factor that binds to heparin and inhibits thrombin-mediated clevage of fibrinogen. Heparin binding, which occurs at ε-amino groups of lysine residues in the protein, is critical to the facilitated function of antithrombin III. It is thus not surprising that in vitro nonenzymatic glycosylation of antithrombin III, which would be expected to render lysine amino groups unavailable, results in a significant diminution of its thrombin-inhibitory activity.[57] Antithrombin III activity has been reported to be

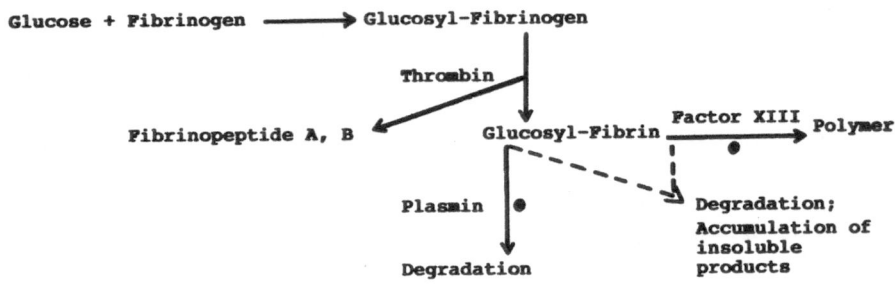

● Compromised by nonenzymatic glycosylation

FIGURE 5-4   Proposed pathophysiologic consequences of nonenzymatic glycosylation of fibrinogen/fibrin.

decreased in patients with insulin-dependent and insulin-independent diabetes,[58,59] and augmented thrombin activity in diabetic subjects is documented by increased plasma and urine levels of fibrinopeptide A.[60] Since the extent of reduction in activity correlated with fasting serum glucose concentrations and glycohemoglobin levels, it appears that in vivo nonenzymatic glycosylation of this protein not only could occur, but also could affect its function, presumably by virtue of an inhibition of the heparin-catalyzed enhancement of thrombin-inhibitory activity. In this light, it is of interest to note that the rate of disappearance of fibrinogen is accelerated in patients with uncontrolled diabetes, and that control of the hyperglycemia normalizes fibrinogen disappearance curves.[61] Decreased antithrombin III activity would further enhance the tendency toward fibrin accumulation that accompanies reduced plasmin-mediated fibrinolysis. However, the reduced survival of fibrinogen in diabetic patients was not believed to be associated with an alteration in the protein,[60] and this cast some doubt on the role of nonenzymatic glycosylation in this change. Furthermore, the thrombin-clotting times of glycosylated and non-glycosylated fibrinogen are reportedly identical,[62] but the results of that study are open to question since glycosylation was performed in vitro, using [^{14}C]glucose and resulting in covalent binding of the radiolabel to the protein. Since aspirin did not alter the extent of radiolabeling, and since impurities in radioactive glucose preparations can covalently bind to proteins, it is likely that significant nonenzymatic glucosylation of lysine residues was not achieved. The radioactive impurities in commerical radiolabeled glucose preparations can give rise to substantial overestimation of the extent of glycosylation if incorporation of radioactivity is the sole criterion used to assess nonenzymatic glycosylation.[63]

Another manner by which glycosylation can impact on hemostasis relates to platelet function, an area that has been extensively studied for many years. Although there is little information as to whether any of the changes described in diabetes, such as increased platelet adhesiveness and aggregation, derive from nonenzymatic glycosylation of platelet proteins, there is evidence that glycosylation of collagen increases its aggregating potency. In one study, collagen extracted from placentas of diabetic patients showed enhanced platelet-aggregating properties compared to that observed with collagen from placentas of nondiabetic patients.[64] An acid-soluble noncollagenous glycoprotein extracted from these placentas also showed increased aggregating potential in diabetic preparations. The level of nonenzymatic glycosylation of both the collagenous and the noncollagenous placental proteins was increased in the diabetic samples, suggesting that the increased aggregation of normal platelets by these proteins results from glucose attachment to lysine residues. A second study examined the aggregating potency of tendon (Type I) collagen prepared from control versus streptozotocin-diabetic rats and of normal tendon collagen subjected to nonenzymatic glycosylation by incubation with glucose in vitro[65] (Figure 5-5). The velocity and intensity of aggregation of normal platelets were increased compared to controls when glycosylated collagen was employed. No effect on the binding of fibrinogen to platelets was observed as a result of glycosylation of this protein.[33]

Enzymes and Hormones

In vitro glycosylation of insulin has been reported, again with findings of potential biomedical significance.[66] Glucosylation diminished the hormone's effect on the oxidation of [^{14}C]glucose by adipose tissue, reduced its ability to stimulate lipogenesis in isolated adipocytes, and decreased its antilipolytic activity. However, since insulin has a relatively short half-life, it is not clear whether in vivo glycosylation of the hormone, if it occurs, would be of sufficient extent to impact on biologic function.

In view of the finding that modification by acetylation of lysine residues in transferrin diminishes this protein's iron-binding capacity and binding to its receptor on reticulocytes,[67] an effect of non-enzymatic glycosylation of transferrin on these parameters might be

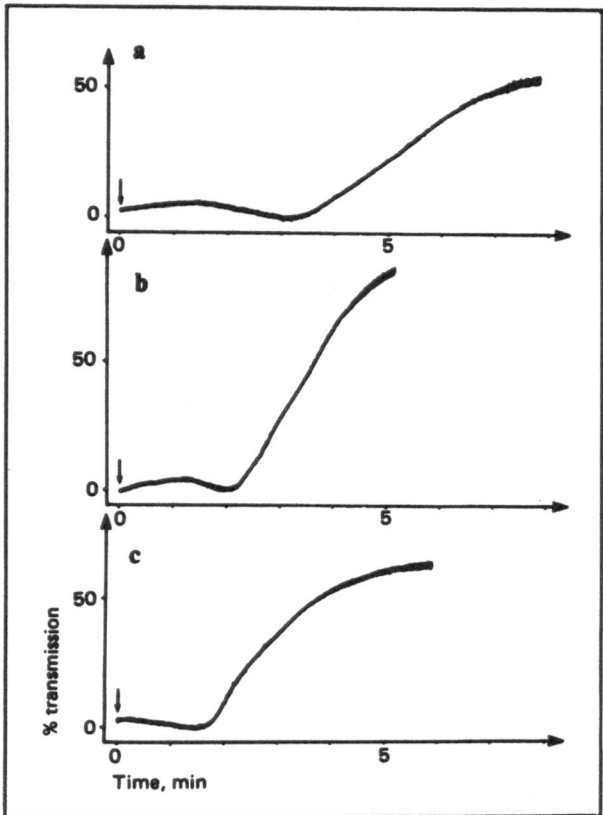

FIGURE 5-5 Aggregation of normal human platelets in the presence of tail tendon collagen from (a) control rats, (b) streptozotocin-diabetic rats, and (c) control rats after in vitro nonenzymatic glycosylation. Reprinted with permission from LePape A et al: Nonenzymatic glycosylation of collagen in diabetes. Incidence on increased normal platelet aggregation. *Haemostasis* 1983;13:36–54.

anticipated. However, no such effect was observed when iron binding by transferrin and specific transferrin binding to Wil-2 cells were compared using nonglycosylated and glycosylated forms of the protein.[33]

Another interesting study found that the nonenzymatic glycosylation of cathepsin B, an enzyme involved in proinsulin cleavage in pancreatic beta cells, partially inhibits the ability of the enzyme to convert proinsulin to insulin in vitro.[68] The suggestion that in vivo glycosylation of pancreatic cathepsin B, if it occurs, could further compromise residual beta cell secretory activity during hyperglycemic episodes in insulin-dependent diabetes is interesting but unproved.

Activity of the renal enzyme N-acetyl-D-glucosaminidase is reduced as a result of in vitro nonenzymatic glycosylation, which causes inactivation by conversion to an isoenzyme.[69] This change in physicochemical and catalytic properties was proportional to the time of incubation and glucose concentration employed and, hence, presumably to the extent of nonenzymatic glycosylation. The enzyme participates in the catabolism of glycosaminoglycans, which are integral constituents of connective tissues and extra-cellular matrices. Its activity is reduced in kidney and other tissues in diabetes, prompting the tempting speculation that this reduction relates to in vivo nonenzymatic glycosylation and consequent inactivation of the enzyme. However, since the glycosaminoglycan content of basement membranes is reportedly decreased rather than increased in diabetes,[70-72] it is difficult to incorporate reduced N-acetyl-glucosaminidase activity due to non-enzymatic glycosylation into a general scheme explaining the pathogenesis of diabetic basement membrane lesions. Similarly, it is not known whether the decreased serum level of two proteinase inhibitors (α_1-inhibitor and α_1-macroglobulin), which correlates with increased nonenzymatic glycosylation of these proteins in diabetic rats, is in some way linked to excess glycosylation.[73]

Tissue and Structural Proteins

Lens Crystallins

The effect of nonenzymatic glycosylation on structural properties of proteins has been best studied in lens, which was one of the first tissues examined in detail. It was also the first tissue in which experimental evidence emerged to invoke the reaction as pathogenetically linked to a defined complication, namely, cataracts. Glucose uptake in the lens is insulin independent, as it is in red blood cells; hence glucose concentration in this tissue parallels that of the extracellular milieu, and lenticular proteins are readily exposed to increased amounts of glucose if hyperglycemia prevails. Once formed, the lens crystalline proteins have little or no turnover, allowing products of nonenzymatic glycosylation, itself an irreversible reaction, to accumulate. In initial experiments, Cerami and associates demonstrated that the ε-amino groups of lysine residues in bovine and rat lens crystallins became glycosylated when incubated in vitro with glucose (50 mM) or glucose-6-phosphate (5 mM), and that under these conditions the lens also developed an opalescence reminiscent of the cloudy cataracts of diabetes.[74,75] The greatest extent of glycosylation, measured as [^{14}C]hexose incorporation, seemed to involve the α-crystallins, although carbohydrate was found associated with all of the crystallins. Increased glycosylation of lens crystallins in the presence of hyperglycemia also occurs in vivo, as demonstrated by increased radioactivity in borohydride-reduced crystallins of lenses removed from rats with alloxan diabetes of 8 to 10 weeks duration and in which cataracts had appeared. The most striking increase in radioactivity was found in the crystallins, and subsequent experiments with cataractous lenses from diabetic and galactosemic rats confirmed that glycosylation in vivo occurred at the ε-amino groups of lysine residues.[75,76] The amount of glucitollysine in lens protein also increases with aging,[77,78] and there is increased glycosylation of lens cortical, but not nuclear, proteins in senile cataracts from diabetic patients.[79,80]

Glycosylation appears to increase the proteins' susceptibility to sulfhydryl oxidation, which in turn leads to the formation of high-molecular-weight aggregates. This interpretation is consistent with the findings that the opalescent crystallins clarify upon addition of reducing agents (borohydride, dithioerythritol, β-mercaptoethanol), and the amount of material eluting in the high-molecular-weight region on gel filtration decreases substantially if the glycosylated crystallins are reduced with dithioerythritol before chromatography (Figure 5-6). Additionally, reduction of proteins prepared from cataractous lenses of galactosemic rats decreases their spectrophotometric absorption (Figure 5-7).

Since neutralization of a charged lysine residue can create conformational charges in α-crystallins that expose inaccessible sulfhydryl groups, it is believed that nonenzymatic glycosylation exerts a similar effect, altering the tertiary structure of the protein and unmasking hidden sulfhydryl groups, which thereby become susceptible to oxidation and can form disulfide cross-links. Information concerning the chemistry of the α-crystallin and the effect of other modifications of its protein amino groups corresponds with this hypothesis. α-crystallin is a polymeric structural protein of lens and consists of two main types of subunits, the A_2 and the B_2. In bovine lens α-crystallin, the A_2 chain has seven lysine residues and one thiol group per subunit, whereas the B_2 chain has 10 lysine residues and no thiol groups per subunit. Carbamylation, like nonenzymatic glycosylation, of lysine amino groups of crystallins with isocyanic acid alters the proteins' surface charge. Carbamylation leads to conformational changes that affect secondary and tertiary structure, increase thiol reactivity, and result in the formation of interchain disulfide bonds.[81] Nonenzymatic glycosylation of α-crystallin also changes the tertiary (but not the secondary) structure and exposes the sulfhydryl goups to a less

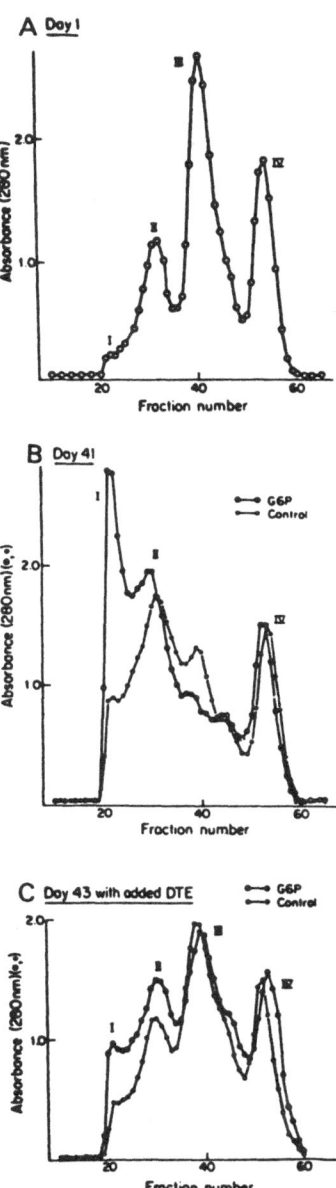

FIGURE 5-6   Effect of nonenzymatic glycosylation of lens crystallins on their gel filtration properties. G6P, glucose-6-phosphate; DTE, dithioerythritol. Reproduced by permission from Cerami A et al: Role of nonenzymatic glycosylation in the development of the sequelae of diabetes mellitus. *Metabolism* 1979;28:431–437.

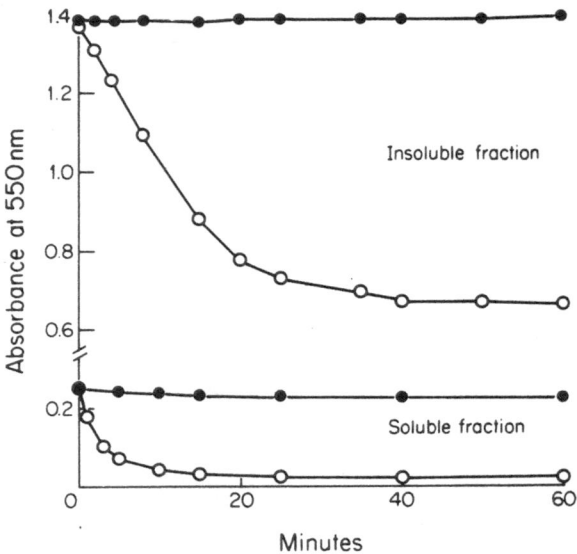

FIGURE 5-7   Effect of reduction with dithioerythritol (DTE) on the absorbance of the soluble and insoluble fractions prepared from cataractous lenses of galactosemic rats. ●, control; O, incubated with DTE. Reproduced from *The Journal of Experimental Medicine*, 1979;150:1098–1107, by copyright permission of The Rockefeller University Press.

hydrophobic microenvironment.[82,83] This is compatible with a partial unfolding of the protein and a facilitation of sulfhydryl oxidation. However, increased sulfhydryl oxidation and disulfide bonding in crystallins from diabetic lenses has also been attributed to a loss of glutathione,[84] possibly related to accumulation of sorbitol as a result of enhanced activity of the polyol pathway.[85] Thus, the exact contribution of nonenzymatic glycosylation to increased sulfhydryl oxidation relative to other metabolic abnormalities in the diabetic lens is not entirely clear. Similarly, the actual role of nonenzymatic glycosylation in the development of cataracts in diabetic patients is still open to question in view of a report that treatment with Sorbinil, an aldose reductase inhibitor, prevented cataracts in galactosemic rats without altering the hexitollysine content of the lens proteins.[86] Furthermore, two studies found no significant correlation between the extent of glycosylation and cataract formation in human and galactosemic rat lenses.[78,87]

    If glycosylation is a causative factor in cataractogenesis, it has been suggested that aspirin might have a beneficial effect by virtue of its ability to acetylate free ε-amino groups of crystallins and other proteins.[88-90] This hypothesis proposes that acetylation of reactive free amino groups makes them unavailable for nonenzymatic glycosylation, and therby prevents crystallin aggregation. In this context, it is of interest to note the report that the prevalence of cataracts in patients with rheumatoid arthritis receiving aspirin was significantly lower than that in a matched population not receiving aspirin.[91] Aspirin has also been reported to prevent the glucose-induced rise in nonenzymtic glycosylation and associated increase in thermal rupture time of collagen when added to in vitro incubations.[90] (see pp. 89, 90).

Nerve

Proteins in sciatic nerves from alloxan-diabetic rats and in femoral nerves from pancreatectomized or alloxan-diabetic dogs have increased nonenzymatic glycosylation compared to nerve proteins in nondiabetic animals.[92] This was demonstrated by subjecting nerve homogenates to reduction with tritiated borohydride, followed by hydrolysis in 6 N hydrochloric acid and affinity chromatography on immobilized m-phenylboronic acid, with expression of the extent of glycosylation as counts per minute per micromole of amino acid. The validity of the technique derives from the ability of phenylboronic acid to complex with cis-dol groups of glycosylated amino acids (see Chapter 3) and the stability of nonenzymatically glycosylated residues to acid hydrolysis after borohydride reduction. There was a threefold increase in the mean level of glycosylated amino acids in peripheral nerve homogenates from diabetic animals compared to samples from nondiabetic animals, with no overlap of individual values between the two groups. The exact nature of the proteins undergoing excess nonenzymatic glycosylation in vivo was not identified, although they were believed to consist of axonal and/or myelin proteins. This study, like those concerning glomerular or retinal microvessel basement membranes, afforded firm evidence that diabetes produces increased nonenzymatic glycosylation of components in a tissue characteristically involved with a chronic diabetic complication. Increased nonenzymatic glycosylation of peripheral nerve from human diabetic subjects has also been found.[93] Subsequent work identified myelin as the principal peripheral nerve constituent that becomes excessively glycosylated and demonstrated analogous findings in central nervous system myelin.[94] A 28,000-dalton peptide in peripheral nerve myelin appears to be the major species undergoing increased glycosylation in vivo, whereas much of the increased radioactivity after reduction with tritiated borohydride of brain myelin eluted in a high-molecular-weight region. Since the involved peripheral nerve myelin peptide is the main structural protein of the myelin sheath, the postulate that nonenzymatic glycosylation might alter the latter's integrity and/or the functional properties of the nerve seems logical. The excessively glycosylated high-molecular-weight species of central nervous system myelin could represent cross-linked proteins derived from glucose-glucose or glucose-protein interactions between glycosylated residues of different peptides (see section on "Advanced Glycosylation End Products").

An important insight into the mechanism by which excess non-enzymatic glycosylation could contribute to the functional and pathologic lesions of diabetic neuropathy was provided by the demonstration that macrophages specifically recognize glycosylated myelin.[95] This recognition leads to endocytosis and intracellular accumulation of myelin, analogous to the receptor-mediated macro-phage uptake of chemically or naturally modified LDLs (see earlier section on "Blood Components"). Intracellular accumulation of myelin from diabetic patients is 2 to 3 times higher than that for myelin from nondiabetic individuals of the same age.[96] Further, receptor interaction of modified proteins may stimulate secretion of proteases such as plasminogen activator; the susceptibility of myelin proteins to plasmin underscores the deleterious potential of such interactions.[97,98] Since segmental demyelination is a characteristic feature of diabetic neuropathy, this process of macrophage recognition could have cogent pathophysiologic significance, linking nonenzy-matic glycosylation to described pathologic changes. Advanced glycosylation end products, formed by dehydration and rearrange-ment of glycosylated proteins, rather than ketoamine adducts, appear to be the recognized forms since exposure of myelin to glucose for 8 weeks was required to demonstrate enhanced macrophage uptake. Similarly, macrophage uptake of myelin isolated from rats with alloxan diabetes of 4 to 6 weeks duration was low, and the rate of accumulation was not different from that observed with myelin prepared from control animals, whereas macrophage uptake of myelin from rats with diabetes of 18 to 24 months duration was greatly exaggerated.

Another manner in which nonenzymatic glycosylation could affect the nervous system relates to its structural and functional conse-quences on tubulin, a microtubular component that participates in neurosecretion and axonal transport. In vitro glycosylation of rat brain tubulin inhibits its ability to polymerize and results in the formation of high-molecular-weight aggregates that persist in the presence of reducing agents.[99] Excess glycosylation of tubulin apparently occurs in vivo in the presence of hyperglycemia, since the amount[3H]boro-hydride-reducible sugar in brain tubulin prepared from diabetic rats is increased. Further, polymerization of tubulin isolated from diabetic rat brains is reduced compared to controls. The fact that lysine residues are required for tubulin polymerization supports the interpretation that nonenzymatic glycosylation of critical lysine groups is responsible for this effect.

Collagen and Basement Membranes

Collagen is the main fibrous protein of connective tissues and is a principal constituent of basement membranes. Collagenous proteins are rich in lysine and hydroxylsine, generally have a long biologic half-life, and are continuously exposed to ambient levels of glucose in the vascular compartment and extracellular fluids. Since variables such as the number of free amino groups in and the residence time of a protein determine the extent of glycosylation in vivo, there are a priori reasons to expect that collagen would be highly subject to excess nonenzymatic glycosylation in vivo. Examination of several collagens has confirmed that this is the case, and a two- to threefold increase has been consistently found when nonenzymatic glycosylation of collagens from tissues of diabetic subjects or animals is compared to that in control samples. This includes aortic, skin, tendon, and glomerular and lens capsule basement membrane collagen[93,100-114] (Figure 5-8). Type I and Type II collagens and glomerular basement membrane also readily undergo nonenzymatic glycosylation in vitro in a reaction that is dependent on time and glucose concentration, although the extent and sites of glucosylation may differ in different collagens.[115-118]

One of the effects of nonenzymatic glycosylation on this class of proteins may be to confer a resistance to collagenase digestion analogous to that occurring with age,[100,101,119,120] but this has not been uniformly found by all investigators. For example, Schnider and Kohn reported that the amount of diaphragm tendon collagen solubilized after 1 hour of collagenase treatment incrased from 2% to 20% over an age range of 18 to 76 years, although the fraction of original ketoamine-linked glucose remaining in the undigested residue did not show any consistent age-associated differences. On the other hand, they found a significant age-related decrease in the acid-soluble and pepsin-released proportions of skin collagen, and an associated increase in the amount of ketoamine-linked glucose in the insoluble fraction with age (Figure 5-9). Findings were similar in skin collagen obtained from three diabetic patients, with a decrease in susceptibility to pepsin digestion, an increase in the proportion of insoluble collagen relative to that expected for their ages, and an increase in the amount of glucose bound to insoluble collagen relative to that in the insoluble fraction of samples from nondiabetic people of comparable ages.

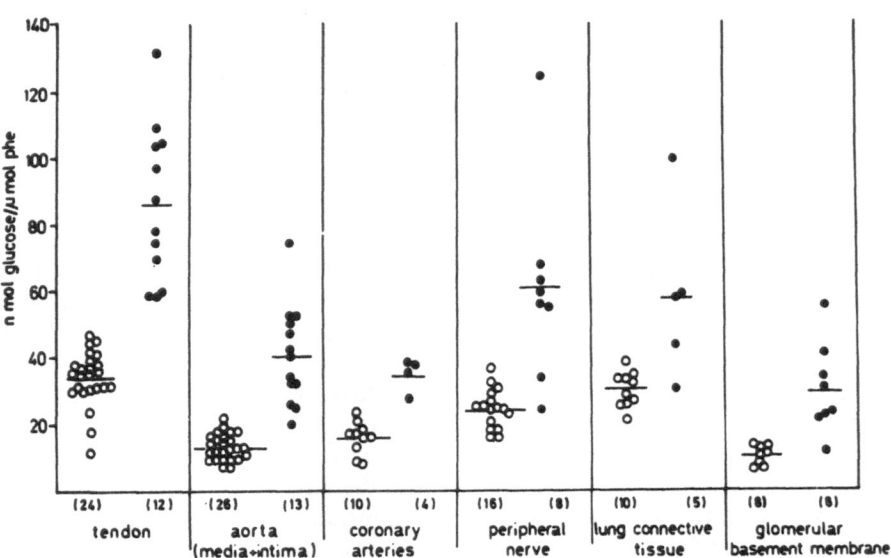

FIGURE 5-8   Glucose bound nonenzymatically to lysine residues of proteins in various tissues from nondiabetic (O) and diabetic (●) subjects. Reproduced with permission of the American Diabetes Association, Inc., from Vogt BW et al: ε-Amino-lysine-bound glucose in human tissues obtained at autopsy: Increase in diabetes mellitus. *Diabetes* 1982;31:1123–1127.

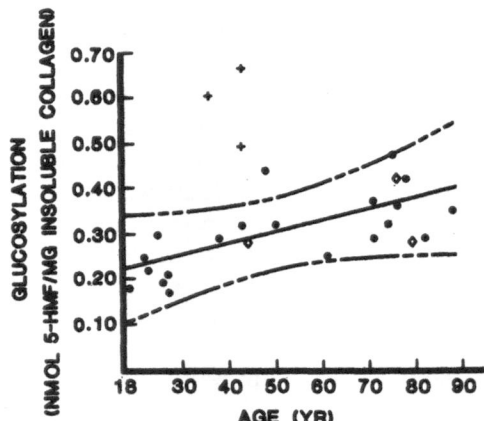

FIGURE 5-9   Levels of nonenzymatic glycosylation (HMF) in insoluble skin collagen from nondiabetic subjects (●), patients with type I diabetes (+), and patients with type II diabetes (◊). Reproduced from *The Journal of Clinical Investigation*, 1981;67:1630–1635, by copyright permission of The American Society for Clinical Investigation.

Another group suggested that decreased degradation of collagen due to excess nonenzymatic glycosylation was the likely mechanism responsible for digital sclerosis in diabetes, although no direct evidence was presented.[121] In contrast, Lyons and Kennedy did not detect an alteration in the susceptibility of normal human skin collagen to collagenase after subjecting the sample to in vitro glycosylation under conditions resulting in a three- to fourfold increase in the amount of ketoamine-linked glucose.[122] Additionally, no difference in the susceptibility of glucosylated Type I collagen to the collagenase from human polymorphonuclear leukocytes is detectable, regardless of the level of glycosylation achieved in vitro.[116]

Interstitial and basement membrane collagens form intermolecular cross-links that play an important role in the stability, maturation, and physicochemical properties of connective tissue fibers and extracellular matrices.[102,123-127] The first step in the generation of these cross-links is an oxidative deamination of certain lysine and hydroxylysine residues that is catalyzed by the enzyme lysyl oxidase.[123,128,129] Reducible intermolecular cross-links can then form by condensation between these lysyl oxidase-generated reactive aldehydes of lysine or hydroxylysine and free ε-amino groups of other lysine or hydroxylysine residues. Since some of these ε-amino groups in collagen and other proteins are also subject to nonenzymatic glycosylation,[107,114,130-132] it is possible that their availability for the enzymatically catalyzed oxidative deamination that is preliminary to aldol condensation is compromised in the face of excess nonenzymatic glycosylation in diabetes (Figure 5-10). At least one study provides experimental evidence to support this postulate.[103] Such a consequence would affect the packing and permeability of basement membranes and could help explain the proteinuria and leaky capillary filtration barriers that are characteristic of diabetes. A recent study examining the influence of nonenzymatic glycosylation on physicochemical properties of Type I collagen offers results that lend credence to this hypothesis.[133] Although molecular structure and stability, assessed by circular dichroism and differential spectrometry, were not affected, viscometric studies demonstrated a lowering in intermolecular interactions and a lessening of hydrophobicity due to addition of hydrophilic glucose groups. When Type I collagen is glycosylated in vitro, glucose attachment occurs mainly on the α_1CB6 peptide, a site participatory in cross-link formation, and the rate of fibril formation is slowed.[134,135] Fibril stabilization via subsequent cross-linking is also reduced. Although the sites of nonenzymatic glucose attachment to collagen molecules in vivo may differ from those to which glucose binds in vitro, it is clear that, at least in vitro, nonenzymatic glycosylation can influence physicochemical properties and packing of collagen fibers.

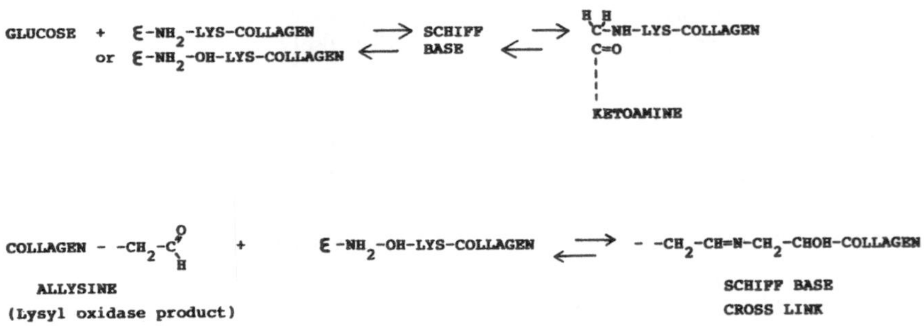

FIGURE 5-10 Similarities between nonenzymatic glycosylation and condensation reaction for cross-link formation.

This is further supported by the observation that glucose, in vitro, inhibits the formation of collagen fibrils.[136] This decrease in fibril formation correlates with a loss in the ability of collagen to serve as a substrate for lysyl oxidase.

An inhibition of normal intermolecular cross-link formation may also help explain the increased solubility of newly synthesized collagen that has been observed when biosynthesis is simultaneously subject to excess nonenzymatic glycosylation. When retinal capillary pericytes are cultured in high glucose concentration, there is an increase in the percentage of total collagen produced by these cells that is in soluble form.[137] Since high glucose concentration did not affect the activity of lysyl oxidase, the increased solubility was ascribed to an inhibition of normal cross-link formation due to occupancy of ε-amino groups of lysine by glucose. It should be noted that the postulate invoking decreased cross-linking on the basis of compromise in the availability of lysine amino groups due to nonenzymatic glycosylation must be taken in the context that collagen from diabetic patients has only a few extra glucose residues per triple-helical molecule.[111] However, the sites of nonenzymatic glycosylation, although limited in number, may be critical to cross-link formation. This would be analogous to the situation with albumin, where modification of even one lysine residue can alter the conformation and binding properties of the protein.[18] It is therefore reasonable to expect that, if particular lysine residues are preferentially glycosylated in vivo, normal collagen cross-linking could be interrupted with only minor degrees of absolute glycosylation.

Certain consequences resulting from an inhibition of normal intermolecular cross-link formation, such as increased solubility, seem contrary to other changes in the properties of collagen that have been described in diabetes and ascribed, in part, to increased nonenzymatic glycosylation. For example, it is difficult to explain decreased susceptibility to collagenase or protease digestion, or increased thermal stability and breaking time of collagen fibers,[89,114,138-140] on the basis of decreased cross-linking, since such findings would, a priori, suggest the opposite. The explanation may relate to the formation of abnormal cross-links or covalent interactions, particularly in the insoluble fraction rather than in newly synthesized collagen. Although the number of sites available for glycosylation may be limited, and the extent of glycosylation stabilizes even if exposure to increased concentrations of glucose continues, rearrangement and dehydration reactions involving the glucoadducts produce compounds capable of interacting with each other and/or with free amino groups of other proteins. This process gives rise to advanced glycosylation end products and abnormal cross-links (discussed extensively in the next section). Additionally, reactive groups formed as a consequence of nonenzymatic glycosylation of collagen can trap, condense, or form covalent bonds with nonglycosylated soluble proteins such as serum albumin, IgG, or lipoproteins (Table 5-1 and Figure 5-11).[141,142] This may explain the increased concentration of albumin in basement membranes of patients with diabetic nephropathy.[143] Further, such trapped or bound proteins retain their ability to form immune complexes with appropriate antibody or antigen, suggesting a mechanism by which nonenzymatic glycosylation could initiate or perpetuate immune injury. Nonenzymatic glycosylation can alter immunogenic properties of proteins, as has been demonstrated with ovalbumin and Type I collagen,[144,145] and thus could, in itself, initiate an immune response directed against components of structural tissues and extracellular matrices. These findings may help explain the characteristic deposition of albumin and IgG that has been detected with immunofluorescent techniques in the microvascular matrix of

TABLE 5-1   Effect of Nonenzymatic Glycosylation on Binding of Proteins to Collagen*

Collagen Preparation	Amount Bound (pmol)	
	Albumin	IgG
Control	45.3 ± 8.1	60.2 ± 13.2
Glycosylated	172.1 ± 13.2	145.0 ± 13.3

*Data from Brownlee et al.[141]

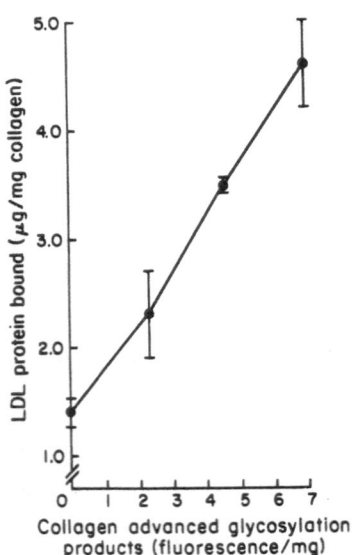

FIGURE 5-11   Binding of LDL to glycosylated collagen as a function of the extent of advanced glycosylation end product formation. Reproduced with permission of the American Diabetes Association, Inc., from Brownlee M et al: Nonenzymatic glycosylation products on collagen covalently trap low-density lipoproteins. *Diabetes* 1985;34:938–941.

tissues from diabetic patients,[24,146,147] and the report that the injection of glycosylated plasma proteins produces a pseudodiabetic thickening of the renal glomerular basement membrane in mice.[23]

It has been proposed that, if physicochemical properties of collagen such as thermal rupture time are altered as a result of nonenzymatic glycosylation, aspirin might have a positive therapeutic effect. When rat tail collagen is incubated with glucose, there are correlative rises in nonenzymatic glycosylation and thermal rupture time, both of which can be prevented when aspirin is added to the incubations.[90] Aspirin given to diabetic animals for 4 weeks also prevented the rise in tail collagen dimerization and thermal rupture time, but did so without affecting the measurable level of nonenzymatic glycosylation of the collagen. The latter finding, coupled with the fact that sodium salicylate was also effective in preventing the rise in thermal rupture time, suggested that mechanisms other than acetylation of lysine residues might be involved in the aspirin effect, and that mechanisms other than nonenzymatic glycosylation might contribute to the change in thermal rupture time in diabetes.

Fibronectin

Fibronectin is one of several noncollagenous extracellular glyco-proteins that interact with cell surfaces and with each other to influence cell growth, development, and adhesion, and to modulate biologic properties of extracellular matrices.[148-150] It is present in large amounts in plasma as a soluble protein, and in an insoluble form at cell surfaces. The two forms are similar but not identical and contain subunits of approximately 220,000 daltons that are joined by disulfide bonds into dimers and multimers. Fibronectin is synthesized by a variety of cells, including fibroblasts and endothelial cells, and it has been demonstrated by immunofluorescent techniques in the con-nective tissue matrix and in basement membranes.[151-153] Although it is not clear whether fibronectin that has been found in glomerular basement membrane derives from local production[154-156] or from entrapment of circulating fibronectin,[157] it is believed that the protein exerts ligand-binding and biologic activities in this matrix that are similar to those it intrinsically possesses and exhibits in other tissue sites.[156] This includes cell-cell and cell-substratum adhesion, binding to collagen and glycosaminoglycans, cell movement, and attachment of cells to extracellular matrix. Indeed, the results of one recent study directly suggest a role for fibronectin in glomerular cell adhesion.[158] The ability of fibronectin to interact with ligands and cell surfaces derives from specific domains of the protein that bind collagen, heparin, heparan sulfate, hyaluronate, fibrin, and plasma membrane recognition sites. Since lysine residues appear necessary for the binding of fibronectin to gelatin (denatured collagen), and probably to the negatively charged glycosaminoglycans as well, occupancy of their ε-amino groups by glucoadducts formed as a result of nonenzymatic glycosylation might be expected to interfere with fibronectin's capacity to interact with these ligands.

Like other proteins, fibronectin is subject to nonenzymatic gly-cosylation in vitro in a reaction that is dependent on time and glucose concentration.[159,160] High levels of glycosylation can be achieved, which result in the formation of fibronectin dimers and multimers (Figure 5-12) and completely inhibit the ability of fibronectin to bind to affinity columns of heparin-Sepharose and gelatin-Sepharose. Glycosylation at a level of 10 times that of native fibronectin also inhibits its binding to heparin when assessed by an in vitro binding assay in which the complexation product formed at neutral pH between fibronectin and [³H]heparin is retained on nitrocellulose filters[159,161] (Figure 5-13). Although one study found that lower levels of glycosylation, about three times those of control, produced a lesser but still significant inhibition of heparin binding in the nitrocellulose assay,[159] another reported that this level of glycosylation did not affect the binding of

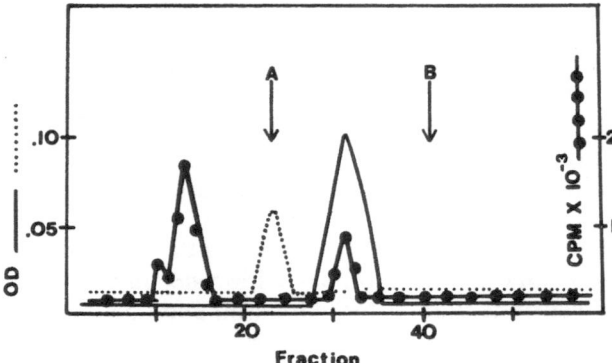

FIGURE 5-12   Gel filtration of native and glucosylated fibronectin. Arrows indicate elution positions of standards with molecular weights of 400,000 (A) and 65,000 (B). Solid curve, fibronectin momomer; ●, highly glycosylated fibronectin; dotted curve, moderately glycosylated fibronectin (15 nmol HMF/ mg protein). Reproduced with permission of the American Diabetes Association, Inc., from Cohen MP, Ku L: Inhibition of fibronectin binding to matrix components by nonenzymatic glycosylation. *Diabetes* 1984;33:970–974.

heparin in solution to human plasma fibronectin in this assay.[160] In the latter study, however, a similar level of glycosylation severely compromised the expected increase in [³H]heparin binding to fibronectin that occurs in the presence of gelatin. A similar decrease was observed when glycosylated gelatin was incubated with [³H]heparin and native fibronectin. Additionally, in vitro glycosylation of fibronectin resulted in a marked decrease in its ability to bind to gelatin-coated netrocellulose filters. Thus, modest levels of glycosylation of fibronectin produce demonstrable alterations in the ligand-binding properties of this glycoprotein. Such levels are comparable to those found in fibronectin isolated from the plasma of alloxan-diabetic dogs and from patients with uncontrolled diabetes. The extent of glycosylation in samples from diabetic subjects who had blood glucose concentrations in excess of 400 mg/dL for several days is about 2 to 3 times that in fibronectin isolated from nondiabetic individuals; nonenzymatic glycosylation of fibronectin in diabetic dogs is about 2½ times that in control dogs, and the extent of nonenzymatic glycosylation in the former group is proportional to the blood glucose concentration.

These studies prompted several postulates concerning the impact of

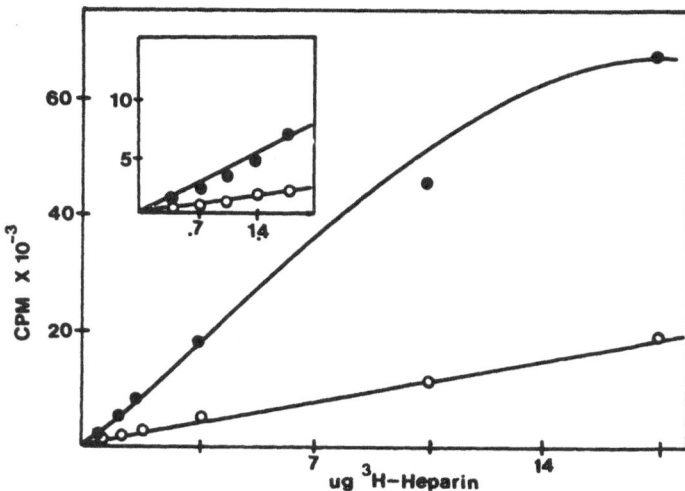

FIGURE 5-13 Binding of native and glucosylated fibronectin to [³H] heparin, assessed by nitrocellulose filter assay. ●, control; ○, glycosylated fibronectin (9 nmol/mg protein). Reproduced with permission of the American Diabetes Association, Inc., from Cohen MP, Ku L: Inhibition of fibronectin binding to matrix components by nonenzymatic glycosylation. *Diabetes* 1984;33:970–974.

nonenzymatic glycosylation of fibronectin on the pathogenesis of diabetic microangiopathy, particularly nephropathy. Since basement membranes contain collagenous proteins and glycosaminoglycans, of which heparan sulfate is the principal species, inhibition of fibronectin binding to these components as a result of excess nonenzymatic glycosylation might influence organizational features of the basement membrane in diabetes. Similarly, the integrity of the microvascular matrix with respect to either self-adhesion or its attachment to adjacent cell surfaces could be compromised by virtue of the inhibitory effect of nonenzymatic glycosylation on the adhesive functions of fibronectin. Additionally, if the extracellular matrix is formed, as suggested, through self-assembly arising from multiple interactions of its macromolecular components, nonenzymatic glycosylation of one or more of these components could perturb this process. This possibility is suggested by the finding that glycosylated fibronectin and gelatin bind poorly to each other and to heparin. Finally, once fibronectin is incorporated into the extracellular matrix at tissue sites,[162] it may have a long half-life, undergo progessive glycosylation, and participate in the formation of advanced glycosylation end products, having further impact on the proteins' matrix-binding properties and altering basement membrane structure.

Advanced Glycosylation End Products

Formation and Identification

The observation that yellow-brown pigments and fluorescent products accumulate with aging and diabetes in tissue proteins with long half-lives such as collagen and lens crystallins, and particularly cataractous lens, suggested the presence of compounds resembling those responsible for the color that appears as a consequence of the nonenzymatic browning reaction in baked or stored foodstuffs[163-166] Such compounds arise from a series of rearrangements and dehydration reactions involving the stable glucose-amino adduct l-deoxyfructosyl lysine and occurring slowly over a period of time. This process creates glucose-derived protein cross-links that involve the addition of other amino groups to form di- and multisubstituted sugars and the condensation of protein-linked glucose molecules through their substituted lysine amino groups.[167-169]

Yellow pigments form after incubation of bovine lens crystallins for many months with glucose or glucose-6-phosphate, and a unique pattern appears in the fluorescence-excitation spectra of these proteins.[157] This consists of a broad shoulder at about 330 nm in the absorption spectra (Figure 5-14), and three novel excitation maxima at 360, 400, and 470 nm (Figure 5-15). The fluorescence-excitation spectra of proteins in human lens cataracts showed patterns similar to those of the glucose-incubated bovine lens crystallins, and the spectra in both sets of samples differed from those observed with suitable controls (noncataractous human lens and bovine lens incubated without hexose). Several fluorescent compounds, co-chromatographing with browning products formed after reaction of glucose with lysine, were also identified in the borohydride-reduced hydrolysate of a cataractous lens.[170,171] Examination of the glucose-incubated lens crystallins by gel-filtration chromatography revealed persistent high-molecular-weight aggregates despite the presence of reducing agent. Subsequent detailed studies confirmed these findings. Four fluorescent, yellow, radioactive peaks were identified on gel filtration of pronase-digested, [³H]borohydride-reduced, acid-hydrolyzed lenses.[172] Amino-acid analysis and HPLC characterized these compounds as lysine derivatives representing browning products. Thus, nonenzymatic glycosylation of lens proteins gives rise to covalent cross-links that are nondisulfide in

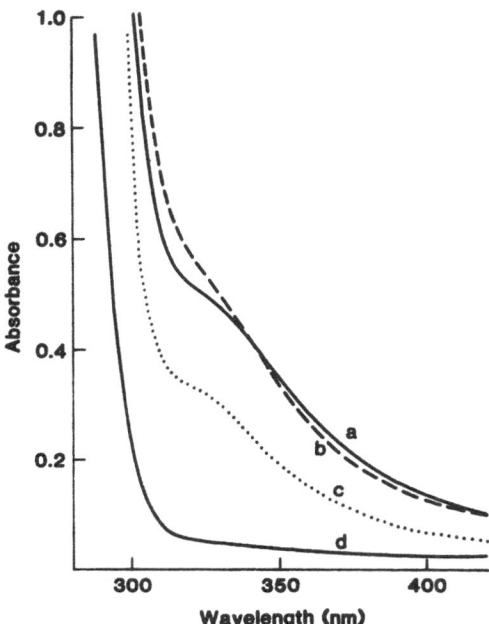

FIGURE 5-14 Absorption spectra of lens protein digests. a, Proteins from cataractous human lens; b, bovine lens proteins incubated with glucose-6-phosphate; c, bovine lens proteins incubated with glucose; d, bovine lens proteins incubated without hexose. Reprinted with permission from Monnier VM, Cerami A: Nonenzymatic browning in vivo: Possible process for aging of long-lived proteins. *Science* 1981;211:491–493. Copyright 1981 by the American Association for the Advancement of Science.

nature, as well as the disulfide cross-links that result from enhanced susceptibility to sulfhydryl oxidation (see pp. 80–82). The nondisulfide cross-links are believed to represent late nonenzymatic browning reaction products.

Formation of pigmented product in collagen as a result of non-enzymatic glycosylation has also been demonstrated. After collagen fibers from rat tail tendon are incubated for about 3 weeks with reducing sugars, the absorption spectra exhibit increases in the range of 300 to 450 nm, compatible with the presence of several chromophores. The observed change was greatest when collagen was incubated with ribose, consistent with the greater glycosylation efficiency of this sugar, and correlated with an increase in the breaking time of the collagen fibers, indicating the formation of cross-links as a result of glycosylation.[139] Of particular interest in this study was the finding that

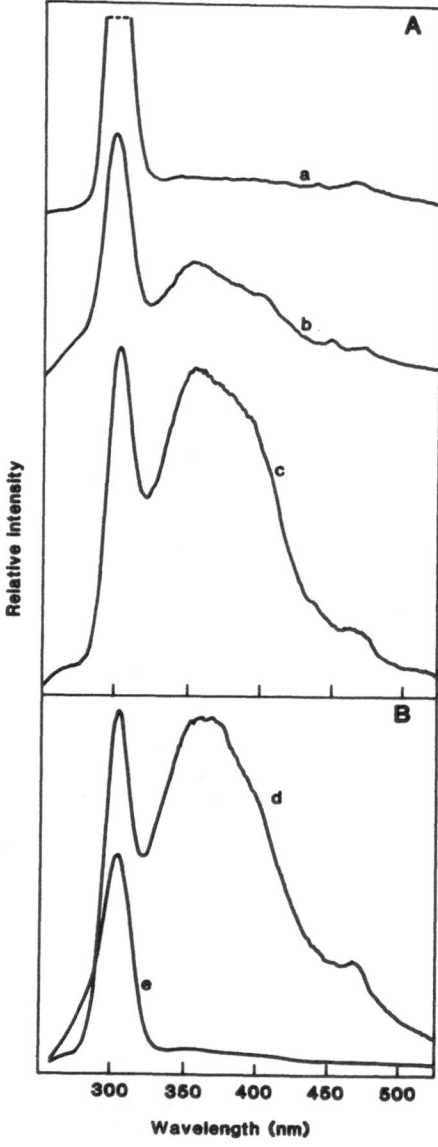

FIGURE 5-15 Excitation spectra of lens protein digests. a, Bovine lens proteins incubated without hexose; b, bovine lens proteins incubated with glucose; c, bovine lens proteins incubated with glucose-6-phosphate; d, proteins from human cataractous lens; e, proteins from young normal human lens. Reprinted with permission from Monnier VM, Cerami A: Nonenzymatic browning in vivo: Possible process for aging of long-lived proteins. *Science* 1981;211:491–493. Copyright 1981 by the American Association for the Advancement of Science.

the level of glycosylation increased during the first 6 days of incubation and then plateaued, whereas changes in absorption and fluorescence continued to increase with time between 3 and 20 days after the start of the incubations (Figure 5-16). The lag time of about 3 days that elapsed before the formation of yellow chromophores is consistent with the notion that these compounds are rearrangement products of the nonenzymatic glycosylation reaction. It is also compatible with the hypothesis that, although the extent of Amadori product formation is dependent on glucose concentration, once formed these products can continue to undergo rearrangements and interactions that are independent of glucose concentration.

Evidence for glucose-medicated covalent cross-linking of collagen after its glycosylation in vitro has also been obtained with rabbit forelimb tendons.[173] Glycosylation was accompanied by a decrease in the solubility of intact tendons and an increase in the force generated on thermal contraction, consistent with the formation of stabilizing covalent bonds. Polyacrylamide gel electrophoresis analysis of cyanogen-bromide digests prepared from glycosylated tendons con-

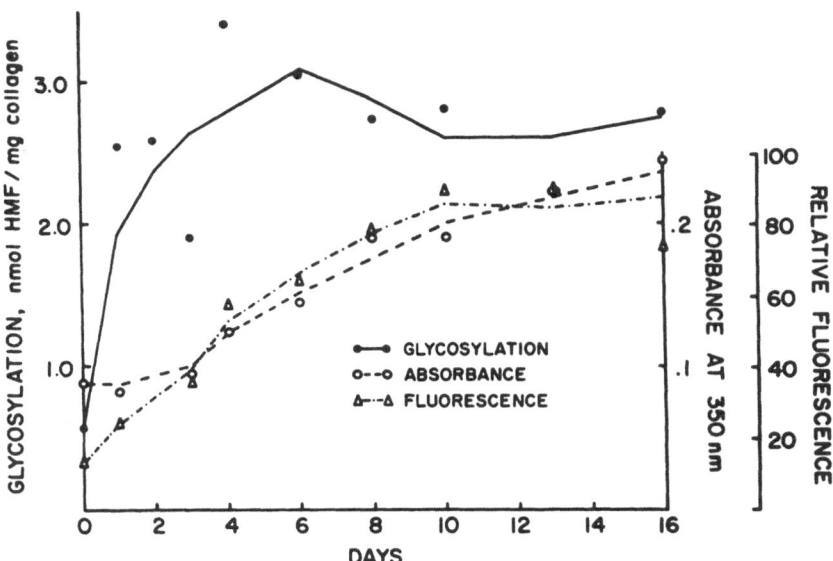

FIGURE 5-16 Time study of levels of nonenzymatic glycosylation (HMF) and amounts of fluorescent and pigmented compounds in digests of rat tail tendon collagen. Reproduced with permission of the American Diabetes Association, Inc., from Kohn RR et al: Collagen aging in vitro by nonenzymatic glycosylation and browning. *Diabetes* 1984;33:57–59.

firmed the presence of high-molecular-weight peptides not detected in the non-glycosylated samples.

A study of insoluble dura mater collagen obtained at autopsy from patients with insulin-dependent and non-insulin-dependent diabetes and from nondiabetic individuals revealed an age-related increase of pigmented material absorbing at 350 nm and 420 nm, and of fluorescence at 440 nm upon excitation at 370 nm.[174] Samples from individuals with Type I diabetes had an increase in pigmented and fluorescent material relative to that observed in samples from nondiabetic subjects of similar ages. These findings not only suggested that advanced glycosylation end products accumulate in collagen with age, but also that this process, and hence aging of connective tissue proteins, is accelerated in diabetes. Changes in the absorption and excitation spectra of collagen from aged and diabetic individuals were similar and resembled the patterns observed after incubation of lens crystallins or rat tail tendon collagen with glucose. These consisted of a new absorption maximum at about 340 nm and enhanced excitation and fluorescent intensities at 370 and 440 nm.

A major fluorescent chromophore formed with nonenzymatic browning has recently been identified. This compound, 2-(2-furoyl)-4(5)-(2-furanyl)-1H-imidazole (FFI), was isolated after acid hydrolysis of albumin or synthetic polylysine that had been incubated with glucose in physiologic solution at 37°C for 28 days.[175] It represents the product formed between two lysine-derived amine nitrogens, each of which is linked to a glucose molecule that had been in the Amadori arrangement of glucose adducts, and has the configuration of a conjugated system of three aromatic heterocycles (Figure 5-17). Thus, there is now a definite chemical explanation for at least one way in which nonenzymatic glycosylation can promote the formation of cross-links between proteins that would not ordinarily occur. A very elegant recent report describes the development of a radioimmuno-assay for FFI.[176] The procedure employs a derivative of FFI as antigen and allows in situ detection of this advanced glycosylation end product by subjecting the material liberated from proteins after acid hydrolysis or proteinase digestion to measurement by the radioimmunoassay. FFI itself has cytotoxic potential in that it can attach to cell proteins and cause lymphocytic death.[177]

Additional information on the nature and consequences of these glucose-derived cross-links has come from studies of albumin and RNase A as model systems. After in vitro glucosylation, albumin develops a new fluorescence at 430 nm and undergoes polymerization.[178] On SDS-polyacrylamide gel electrophoresis, native albumin migrates as monomer and dimer, whereas a substantial portion of the glycosylated derivatives migrate as aggregates, up to a hexameric form.

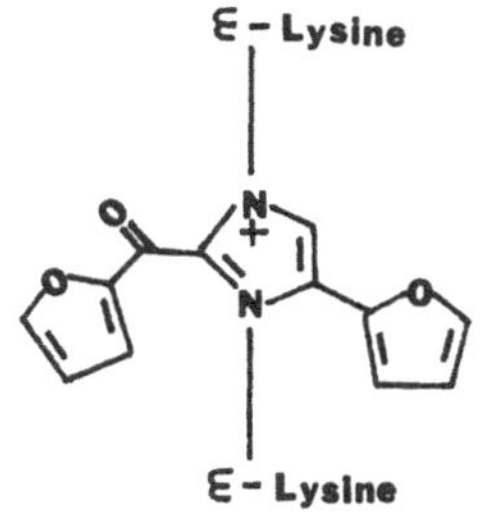

FIGURE 5-17   Structure of condensation product 2-(2-furoyl)-4(5)-(2-furanyl)-1H-imidazole (FFI), formed between two lysine-derived amino groups, each of which is linked to a glucose molecule.

These multimers were demonstrable in the presence of mercaptoethanol, indicating that they represented nondisulfide cross-linking. Glucosylation of RNase A results in a time-dependent formation of RNase dimer and trimer, again demonstrable in the presence of reducing agent.[179] The protein continued to polymerize even after the removal of free glucose, supporting the hypothesis that, once initiated by hyperglycemia, nonenzymatic glycosylation can continue to exert deleterious effects as a result of cross-linking even though a hyperglycemic state no longer prevails. When RNase was radioiodinated and then glycosylated,[125I]-labeled dimer was formed, indicating that either two glycosylated proteins or one glycosylated and one nonglycosylated protein became cross-linked. That dimerization occurred by the latter interaction was demonstrated with the formation of radioactive polymer after incubating nonradioactive glycosylated protein with radioiodinated nonglycosylated protein. This cross-linking between glucosylated and native RNase in the absence of glucose could be inhibited by lysine. Thus, despite one report to the contrary,[180] it appears that a glucosylated lysine residue on one molecule can condense with a free amino group of a lysine residue on another molecule. These results not only provide a chemical basis for another way in which nonenzymatic glycosylation can cross-link proteins, but also offer an explanation for the ability of glycosylated proteins to trap or bind unrelated proteins.

Such advanced glycosylation end products, rather than the Amadori glucose adduct, appear responsible for the recognition of myelin by macrophages.[95] The uptake by mouse peritoneal macrophages of myelin from animals diabetic for 4 to 6 weeks was low and was comparable to that observed with myelin prepared from nondiabetic animals. In contrast, macrophage accumulation of myelin from rats with diabetes of 1 to 2 years duration was several times that of myelin from control animals. Macrophage recognition of proteins modified by advanced glycosylation end products may, in fact, provide a biologic mechanism for removal and degradation of long-lived structural macromolecules as they age. Vlassara and colleagues recently reported that bovine serum albumin that has been modified by advanced glycosylation end products is specifically bound, internalized, and degraded at 37°C by mouse peritoneal macrophages in a dose-dependent and saturable manner.[181] These processes are mediated by a macrophage receptor that specifically recognizes advanced glycosylation end product modifications, including FFI. However, macrophages accumulated bovine serum albumin that had been chemically coupled to FFI only one third as efficiently as they took up albumin that was more generally modified by advanced glycosylation end products, suggesting that the receptor has a higher affinity for other advanced glycosylation end products than it does for FFI.

Clinical Correlates

The experimental results discussed above indicate that glucoadducts produced as a result of nonenzymatic glycosylation can give rise to advanced glycosylation end products that have, in themselves, the potential for disturbing structural or biological features of involved proteins and for exerting toxic effects on cell processes. These products can continue to form even after the hyperglycemic stimulus is removed, since, unlike the nonenzymatic glycosylation reaction, their formation is independent of glucose concentration. For long-lived proteins, the impact of abnormal cross-linking between advanced glycosylation end products or their binding of unrelated molecules could be substantial; this concept assumes special significance when one considers that several of the characteristic complications of diabetes occur in sites where component proteins have relatively slow turnover times. Mechanistically, events consequent to the formation of advanced glycosylation end products could provide the missing pathogenetic link that has eluded attempts to correlate diabetic control with the development or progression of chronic complications. In this construct, a period of hyperglycemia would initiate excess nonenzymatic glycosylation, but deleterious effects could ensue as a result of protein interactions and abnormalities engendered by subsequent formation of advanced end products even after the hyperglycemia is corrected. Simply stated, the proverbial horse would be out of the barn door, since glucoadducts, once formed, would provide a chemical framework for self-perpetuating and damaging processes that can continue even if strict diabetic control is instituted. This hypothesis, if correct, provides one of the strongest arguments in favor of early aggressive therapeutic intervention with intensified regimens to establish and maintain normoglycemia. Future studies examining the correlation between selected parameters reflecting diabetic complications, such as basement membrane thickness or albuminuria, at various stages, and the simultaneous amount of glycosylated end products in the relevant tissue, such as glomerular basement membrane, should help corroborate or refute this hypothesis. The recent development of a sensitive and reproducible assay to quantitate at least one of these products in tissue samples of limited size has opened the way for these and other necessary studies, including possible identification of the stage at which incipient glycosylation-related damage might still be reversible.

References

1. Bunn HF, Briehl RW: The interaction of 2,3-diphosphoglycerate with various human hemoglobins. *J Clin Invest* 1970;49:1088–1095.
2. Ditzel J, Anderson H, Peters ND: Oxygen affinity of hemoglobin and red cell 2,3-diphosphoglycerate in childhood diabetes. *Acta Pediatr Scand* 1975;64:355–361.
3. Farris L, Wajcman H, Jones RT, et al: Functional properties of hemoglobin A_{1c}. *Clin Res* 1977;25:115A.
4. McDonald MJ, Bleichman M, Bunn HF, et al: Functional properties of the glycosylated minor components of human adult hemoglobin. *J Biol Chem* 1979;254:702–707.
5. Arturson G, Garby L, Robert M, et al: Oxygen affinity of whole blood in vivo and under standard conditions in control subjects with diabetes mellitus. *Second J Clin Lab Invest* 1974;34:19–22.
6. Ditzel J, Standl E: The problem of tissue oxygenation in diabetes mellitus. *Acta Med Scand* 1978; suppl 578:59–68.
7. Ditzel J, Nielsen NV, Kjaergaard JJ: Hemoglobin A_{1c} and red cell oxygen release capacity in relation to early retinal changes in newly discovered overt and chemical diabetes. *Metabolism* 1979;28(suppl 1):440–447.
8. Ditzel J: Changes in red cell oxygen release capacity in diabetes mellitus. *Fed Proc* 1977;38:2484–2488.
9. Ditzel J: Affinity hypoxia as a pathogenic factor of microangiopathy with particular reference to diabetic retinopathy. *Acta Endocrinol* 1980;94:39–55.
10. Ditzel J, Daugaard P, Anderson H: Oxygen affinity of haemoglobin and red cell 2,3 diphosphoglycerate in childhood diabetes. *Diabetologia* 1974;10:363.
11. Ditzel J, Jaeger P, Standl E: An adverse effect of insulin on the oxygen release capacity of red blood cells in nonacidotic diabetes. *Metabolism* 1978;27:929–934.

12. Standl E, Kolb HJ: 2,3-Diphosphoglycerate fluctuations in erythrocytes reflecting pronounced blood glucose variance: In vivo and in vitro studies in normal, diabetic and hypoglycemic subjects. Diabetologia 1973;9:461–466.

13. Ditzel J, Kawahara R, Mourits-Andersen T, et al: Changes in blood glucose, glycosylated hemoglobin and hemoglobin-oxygen affinity following meals in diabetic children. Eur J Pediatr 1981;137:171–174.

14. Samaja M, Melotti D, Carenini A, et al: Glycosylated haemoglobins and the oxygen affinity of whole blood. Diabetologia 1982;23:399–402.

15. Bunn HF, Gabbay KH, Gallop PM: The glycosylation of hemoglobin: Relevance to diabetes mellitus. Science 1978;200:21–27.

16. Bunn HF: Nonenzymatic glycosylation of protein: Relevance in diabetes Am J Med 1981;70:325–330.

17. Day JF, Thornburg RW, Thorpe SW, et al: Nonenzymatic glucosylation of rat albumin: Studies in vitro and in vivo. J Biol Chem 1979;254:9394–9400.

18. Shaklai N, Garlick RL, bunn HF: Nonenzymatic glycosylation of human serum albumin alters its conformation and function. J Biol Chem 1984;259:3812–3817.

19. Williams SK, Devenney JJ, Bitensky MW: Micropinocytic ingestion of glycosylated albumin by isolated microvessels: Possible role in pathogenesis of diabetic microangiopathy. Proc Natl Acad Sci USA 1981; 78:2393–2397.

20. Williams SK, Solenski NJ: Enhanced vesicular ingestion of nonenzymatically glucosylated proteins by capillary endothelium. Microvasc Res 1984;28:311–321.

21. Williams SK, Siegal RK: Preferential transport of nonenzymatically glucosylated ferritin across the kidney glomerulus. Kidney Int 1985;28: 146–152.

22. Ghiggeri GM, Candiano G, Delfino G, et al: Glycosyl albumin and diabetic microalbuminuria: Demonstration of altered renal handling. Kidney Int 1984;25:565–570.

23. McVerry BA, Hopp A, Fisher C, et al: Production of pseudodiabetic renal glomerular changes in mice after repeated injection of glycosylated proteins. Lancet 1980;2:738–740.

24. Jeraj KP, Michael AF, Mauer SM, et al: Glucosylated and normal human or rat albumin do not bind to renal basement membranes of diabetic and control rats. Diabetes 1983;32:380–382.

25. Miller K, Michael AF: Immunopathology of renal extracellular membranes in diabetes mellitus: Specificity of tubular basement membrane immunoflourescence. Diabetes 1976;25:701–708.

26. Nathke HE, Siess EA, Wieland OH: Glucosylated plasma protein infection does not produce glomerular basement membrane thickening. Horm Metab Res 1984;16:557–558.

27. Summerfield JA, Vergalla J, Jones EA: Modulation of a glycoprotein recognition system on rat hepatic endothelial cells by glucose and diabetes mellitus. J Clin Invest 1982;69:1337–1347.

28. Ashwell G, Morell AG: The role of surface carbohydrates in the hepatic recognition and transport of circulating glycoproteins. *Adv Enzymol* 1974;41:99-128.

29. Achord DT, Brot FE, Sly WS: Inhibition of the rat clearance system for agalacto-orosomucoid by yeast mannans and by mannose. *Biochem Biophys Res Commun* 1977;77:409-415.

30. Achord DT, Bort FE, Bell CE, Sly WS: Human β-glucuronidase: In vivo clearance and in vitro uptake by a glycoprotein recognition system on reticuloendothelial cells. *Cell* 1978;15:269-279.

31. Cohenford MA, Urbanowski JC, Shepard DC, et al: Nonenzymatic glycosylation of human IgG: *In vitro* preparation. *Immunol Commun.* 1983;12:189-200.

32. Urbanowski JC, Cohenford MA, Dain JA: Nonenzymatic galactosylation of human serum albumin. *J Biol Chem* 1982;257:111-115.

33. Ney KA, Pasqua JJ, Colley KJ, et al: In vitro preparation of non-ezymatically glucosylated human transferrin,α_2,and fibrinogen with preservation of function. *Diabetes* 1985;34:462-470.

34. Marynen P, Vanleuven F, Cassiman J-J et al: Proteolysis at a lysine residue abolishes the receptor-recognition site of alpha 2-macroglobulin complexes. *FEBS Lett* 1982;137:241-244.

35. Schleicher E, Deufel T, Wieland OH: Non-enzymatic glycosylation of human serum lipoproteins: Elevated ε-lysine glycosylated low density lipoprotein in diabetic patients. *FEBS Lett* 1981;129:1-4.

36. Sasaki J, Arora V, Cottam GL: Nonenzymatic galactosylation of human LDL decreases its metabolism by human skin fibroblasts. *Biochem Biophys Res Commun* 1982;108:791-796.

37. Witztum JL, Mahoney EM, Branks MJ, et al: Nonenzymatic glucosylation of low-density lipoproteins alters its biologic activity. *Diabetes* 1982;31:283-291.

38. Kim HJ, Kurup IV: Nonenzymatic glycosylation of human plasma low density lipoprotein: Evidence for in vitro and in vivo glycosylation. *Metabolism* 1982;31:348-353.

39. Sasaki J, Cottam GL: Glycosylation of human LDL and its metabolism in human skin fibroblasts. *Biochem biophys Res Comm* 1982;104:977-983.

40. Gonen B, Baenziger J, Schonfeld G, et al: Non-enzymatic glycosylation of low density lipoproteins in vitro: Effects on cell-interactive properties. *Diabetes* 1981;30:875-878.

41. Lorenzi M, Cagliero E, Markey B, et al: Interaction of human endothelial cells with elevated glucose concentrations and native and glycosylated low density lipoproteins. *Diabetologia* 1984;26:218-222.

42. Kim HJ, Kurup IV: Decreased catabolism of glycosylated low density lipoprotein in diabetic rats. *Diabetes* 1981;30:47A.

43. Sasaki J, Cottam GL: Glycosylation of LDL decreases its ability to interact with high-affinity receptors of human fibroblasts in vitro and decreases its clearance from rabbit plasma in vivo. *Biochem Biophys Acta* 1982;713:199-207.

44. Schleicher E, Olgemöller B, Schön J, et al: Limited nonenzymatic glucosylation of low-density lipoprotein does not alter its catabolism in tissue culture. *Biohcem Biophys Acta* 1985;846:226–233.
45. Kramer FB, Chen Y-DI, Cheung RMC, et al: Are the binding and degradation of low density lipoprotein altered in Type 2 (non-insulin dependent) diabetes mellitus? *Diabetologia* 1981;23:28–33.
46. Lopez-Virella MF, Sherer, GK, Lees AM, et al: Surface binding, internalization and degradation by cultured human fibroblasts of low density lipoproteins isolated from Type I (insulin-dependent) diabetic patients: Changes with metabolic control. *Diabetologia* 1982;22:430–436.
47. Steinbrecher VP, Witztum JL: Glucosylation of low-density lipoproteins to an extent comparable to that seen in diabetes slows their catabolism. *Diabetes* 1984;33:130–134.
48. Curtiss LK, Witztum JL: A novel method for generating region-specific monoclonal antibodies to modified proteins: Application to the identification of human glucosylated low-density lipoproteins. *J Clin Invest* 1983;72:1427–1438.
49. Curtiss LK, Witzum JL: Plasma apolipoproteins Al, All, B, Cl and E are glucosylated in hyperglycemic diabetic subjects. *Diabetes* 1985;34:452–461.
50. Kesaniemi YA, Witztum JL, Steinbrecher UP: Receptor-mediated clearance of low density lipoprotein in man: New estimates using glucosylated low density lipoprotein. *Arteriosclerosis* 1982;2:441a.
51. Sasaki J, Okamura T, Cottam GL: Measurement of receptor-independent metabolism of low-density lipoprotein: An application of glycosylated low-density lipoprotein. *Eur J Biochem* 1983;131:535–538.
52. Goldstein JL, Ho YK, Basu SK, et al: Binding site on macrophages that mediates uptake and degradation of acetylated low density lipoprotein, producing massive cholesterol deposition. *Proc Natl Acad Sci USA* 1979;76:333–337.
53. Gralnick HG, Coller BS, Sultan Y: Carbohydrate deficiency of the factor VIII/von Willebrand factor protein in Von Willebrand's disease variants. *Science* 1976;192:56–59.
54. Saraswathi S, Colman RW: Role of galactose in bovine factor V *J Biol Chem* 1975;250:8111–8118.
55. Lutjens A, teVelde AA, v.d. Veen EA, et al: Glycosylation of human fibrinogen in vivo. *Diabetologia* 1985;28:87–89.
56. Brownlee M, Vlassara H, Cerami A: Nonenzymatic glycosylation reduces the suseptibility of fibrin to degradation by plasmin. *Diabetes* 1983; 32:680–684.
57. Brownlee M, Vlassara H, Cerami A: Inhibition of heparin-catalyzed human antithrombin III activity by nonenzymatic glycosylation. *Diabetes* 1984;33:532–535.
58. Banerjee RN, Sahni AL, Kumar V, et al: Antithrombin 3 deficiency in maturity onset diabetes mellitus and atherosclerosis. *Thromb Diath Haemorrh* 1974;31:339–345.
59. Sowers JR, Truck ML, Sowers DK: Plasma antithrombin III and

thrombin generation time: Correlation with hemoglobin A and fasting serum glucose in young diabetic women. *Diabetes Care* 1980;3:655–658.

60. Jones RL: Fibrinopeptide-A in diabetes mellitus: Relation to levels of blood glucose, fibrinogen disappearance, and hemodynamic changes. *Diabetes* 1985;34:836–843.

61. Jones RL, Peterson CM: Reduced fibrinogen survival in diabetes mellitus: A reversible phenomenon. *J Clin Invest* 1979;63:485–493.

62. McVerry BA, Thorpe S, Joe F, et al: Nonenzymatic glucosylation of fibrinogen. *Haemostasis* 1981;10:261–270.

63. Trueb B, Holenstein CG, Fischer RW, et al: Nonenzymatic glycosylation of proteins. A warning. *J Biol Chem* 1980;255:6717–6720.

64. LePape A, Gutman N, Guitton JD, et al: Nonenzymatic glycosylation increases platelet aggregating potency of collagen from placenta of diabetic human beings. *Biochem Biophys Res Commun* 1983;111:602–610.

65. LePape A, Guitton JD, Gutman N, et al: Nonenzymatic glycosylation of collagen in diabetes: Incidence on increased normal platelet aggregation. *Haemostasis* 1983;13:36–54.

66. Dolhofer R, Wieland OH: Preparation and biological properties of glycosylated insulin. *FEBS Lett* 1979;100:133–136.

67. Kornfeld S: The effects of structural modification in the biologic activity of human transferrin. *Biochemistry* 1968;1:945–954.

68. Coradello H, Pollak A, Pagano M, et al: Nonenzymatic glycosylation of Cathepsin B: Possible influence on conversion of proinsulin to insulin. *IRCS Med Sci* 1981;9:766–767.

69. Dolhofer R, Siess EA, Wieland OH: Inactivation of bovine kidney β-*N*-acetyl-D-glucosaminidase by nonenzymatic glucosylation. *Hoppe Seylers Z Physiol Chem* 1982;363:1427–1436.

70. Parathasarathy N, Spiro RG: Effect of diabetes on the glycosaminoglycan component of the human glomerular basement membrane. *Diabetes* 1982;31:738–741.

71. Cohen MP, Surma ML: [^{35}S]-Sulfate incorporation into glomerular basement membrane glycosaminoglycans is decreased in experimental diabetes. *J Lab Clin Med* 1981;98:715–722.

72. Cohen MP, Surma ML: Effect of diabetes on in vivo metabolism of [^{35}S]-labeled glomerular basement membrane. *Diabetes* 1984; 33:8–12.

73. Esnard F, Guitton JD, Stauber WT, et al: Nonenzymatic glycosylation of rat serum proteinase inhibitors and change in their concentration during experimental diabetes. *Molecular Physiology* 1985;7:211–218.

74. Stevens VJ, Rouzer CA, Monnier VM et al: Diabetic cataract formation: Potential role of glycosylation of lens crystallins. *Proc Natl Acad Sci USA* 1978;75:2918–2922.

75. Cerami A, Stevens VJ, Monnier VM: Role of nonenzymatic glycosylation in the development of the sequelae of diabetes mellitus. *Metabolism* 1979;28:431–437.

76. Monnier VM, Stevens VJ, Cerami A: Nonenzymatic glycosylation, sulfhydryl oxidation, and aggregation of lens proteins in experimental sugar cataracts. *J Exp Med* 1979;150:1098–1107.

77. Chiou SH, Chylack LT, Tung WH, et al: Nonenzymatic glycosylation of bovine lens crystallins: Effect of aging. *J Biol Chem* 1981;256:5176–5180.
78. Pande A, Garner WH, Spector A: Glucosylation of human lens protein and cataractogenesis. *Biochem Biophys Res Commun* 1979;89:1260–1266.
79. Lee JH, Skin DH, Lupovitch A, et al: Glycosylation of lens proteins in senile cataract and diabetes mellitus. *Biochem Biophys Res Commun* 1984;123:888–893.
80. Kasai K, Nakamura T, Kase N, et al: Increased glycosylation of proteins from cataractous lenses in diabetes. *Diabetologia* 1983;25:36–38.
81. Beswick HT, Harding JJ: Conformational changes induced in bovine lens α-crystallin by carbamylation: Relevance to cataract. *Biochem J* 1984;223:221–227.
82. Liang JN, Chylack T: Change in the protein tertiary structure with nonenzymatic glycosylation of calf α-crystallin. *Biochem Biophys Res Commun* 1984;123:899–906.
83. Liang JN, Chakrabarti B: Glycosylation-induced conformational change in α-crystallin of bovine lens. *Biophys J* 1981;33:138a.
84. Ansari NHM, Awsathi YL, Srivastava SK: Role of glycosylation in protein disulfide formation and cataractogenesis. *Exp Eye Res* 1980;31:9–19.
85. Kinoshita JH: Mechanisms initiating cataract formation. *Invest Ophthalmol* 1974;13:713–724.
86. Chiou SH, Chylack LT, Bunn HF, et al: Role of nonenzymatic glycosylation in experimental cataract formation. *Biochem Biophys Res Commun* 1980;95:894–901.
87. Ansari NH, Awasthi YC, Srivastava SK: Presented at the Annual Spring Meeting of the Association for Research in Vision and Ophthalmology, Florida, 1979.
88. Rao GN, Lardis MP, Cotlier E: Acetylation of lens crystallins: A possible mechanism by which aspirin could prevent cataract formation. *Biochem Biophys Res Commun* 1985;128:1125–1132.
89. Ceriello A, Dello Russo P, Curcio F, et al: Acetylsalicylic acid and lysine inhibit protein glycosylation in vitro: A preliminary report. *Diabete Metab* 1984;10:128–129.
90. Yue DK, McLennan S, Handelsman DJ, et al: The effect of salicylates on nonenzymatic glycosylation and thermal stability of collagen in diabetic rats. *Diabetes* 1984;33:745–751.
91. Cotlier E: Aspirin effect on cataract formation in patients with rheumatoid arthritis alone or combined with diabetes. *Int Ophthalmol* 1981;3:173–177.
92. Vlassara H, Brownlee M, Cerami A: Nonenzymatic glycosylation of peripheral nerve protein in diabetes mellitus. *Proc Natl Acad Sci USA* 1981;78:5190–5192.
93. Vogt BW, Schleicher ED, Wieland OH: ε-amino-lysine-bound glucose in human tissues obtained at autopsy: Increase in diabetes mellitus. *Diabetes* 1982;31:1123–1127.
94. Vlassara H, Brownlee M, Cerami A: Excessive nonenzymatic glycosy-

lation of peripheral and central nervous system myelin components in diabetic rats. *Diabetes* 1983;32:670–674.

95. Vlassara H, Brownlee M, Cerami A: Accumulation of diabetic rat peripheral nerve myelin by macrophages increases with the presence of advanced glycosylation end-products. *J Exp Med* 1984;160:197–207.

96. Vlassara H, Brownlee M, Cerami A: Recognition and uptake of human diabetic peripheral nerve myelin by macrophages. *Diabetes* 1985;34:553–557.

97. Johnson WJ, Pizzo SV, Imber MJ, et al: Receptors for maleylated proteins regulate secretion of neutral proteases by murine macrophages. *Science* 1982;218:574–576.

98. Cammer W, Brosnan CF, Bloom BR, et al: Degradation of P_0, P_1, and Pr proteins in peripheral nervous system myelin by plasmin: Diseases. *J Neurochem* 1981;36:1506–1514.

99. Williams SK, Howarth NL, Devenny JJ, et al: Structural and functional consequences of increased tubulin glycosylation in diabetes mellitus. *Proc Natl Acad Sci USA* 1982;79:6546–6550.

100. Rosenberg H, Modrak JB, Hassing JM, et al: Glycosylated collagen. *Biochem Biophys Res Commun* 1979;91:498–501.

101. Schnider SL, Kohn RR: Glucosylation of human collagen in aging and diabetes mellitus. *J Clin Invest* 1980;66:1179–1181.

102. Schnider SL, Kohn RR: Effects of age and diabetes mellitus on the solubility and nonenzymatic glucosylation of human skin collagen. *J Clin Invest* 1981;67:1630–1635.

103. LePape A, Guitton JD, Muh JP: Modification of glomerular basement membrane cross-links in experimental diabetic rats. *Biochem Biophys Res Commun* 1981;100:1214–1221.

104. LePape A, Muh JP, Bailey AJ: Characterization of N-glycosylated Type I collagen in streptozotocin-induced diabetes. *Biochem J* 1981;197:405–412.

105. Buckingham BA, Uitto J, Sandberg C, et al: Scleroderma-like changes in insulin-dependent diabetes mellitus: Clinical and biochemical studies. *Diabetes Care* 1984;7:163–169.

106. Chang AY, Noble RE: 5-Hydroxymethylfurfural-forming proteins in the renal glomeruli of control and streptozotocin-diabetic rats. *Life Sci* 1980; 26:1329–1333.

107. Cohen MP, Urdanivia E, Surma M, et al: Increased glycosylation of glomerular basement membrane collagen in diabetes. *Biochem Biophys Res Commun* 1980;95:765–769.

108. Cohen MP, Wu V-Y: Identification of specific amino acids in diabetic glomerular basement membrane collagen subject to nonenzymatic glucosylation in vivo. *Biochem Biophys Res Commun* 1981;100:1549–1544.

109. Perejda AJ, Uitto J: Nonenzymatic glycosylation of collagen and other proteins: Relationship to development of diabetic complications. *Coll Rel Res* 1982;2:81–88.

110. Schleicher E, Wieland OH: Changes of human glomerular basement membrane in diabetes mellitus. *J Clin Chem Biochem* 1984;22:223–227.

111. Trüeb B, Flückiger R, Winterhalter KH: Nonenzymatic glycosylation of

basement membrane collagen in diabetes mellitus. *Coll Relat Res* 1984; 4:239–251.

112. Uitto J, Perejda AJ, Grant GA, et al: Glucosylation of human glomerular basement membrane collagen: Increased content of hexose in ketoamine linkage and unaltered hydroxylysine-O-glycosides in patients with diabetes. *Connect Tissue Res* 1982;10:287–296.

113. Mandel SS, Shin DH, Newman BL, et al: Glycosylation in vivo of human lens capsule (basement membrane) and diabetes mellitus. *Biochem Biophys Res Commun* 1983;117:51–56.

114. Andreassen TT, Seyer-Hansen K, Bailey AJ: Thermal stability, mechanical properties and reducible cross-links of rat tail tendon in experimental diabetes. *Biochim Biophys Acta* 1983;677:313–317.

115. Rogozinski S, Blumenfeld OO, Seifter S: The nonenzymatic glycosylation of collagen. *Arch Biochem Biophys* 1983;221:427–437.

116. Perejda AJ, Zaragoza EJ, Eriksen E, et al: Nonenzymatic glucosylation of lysyl and hydroxylysyl residues in Type I and Type II collagens. *Coll Relat Res* 1984;4:427–439.

117. Cohen MP, Urdanivia E, Wu V-Y: Nonenzymatic glycosylation of basement membrane. *Renal Physiol* 1981;4:90–95.

118. Cohen MP, Urdanivia E, Surma M, et al: Nonenzymatic glycosylation of basement membranes. In vitro studies. *Diabetes* 1981;30:367–371.

119. Hamlin CR, Kohn RR: Evidence for progressive, age-related structural changes in post-mature human collagen. *Biochim Biophys Acta* 1971; 236:458–467.

120. Hamlin CR, Kohn RR, Luschin JH: Apparent accelerated aging of human collagen in diabetes mellitus. *Diabetes* 1975;24:902–904.

121. Seibold JR, Uitto J, Dorwart BB, et al: Collagen synthesis and collagenase activity in dermal fibroblasts from patients with diabetes and digital sclerosis. *J Lab Clin Med* 1985;105:664–667.

122. Lyons TJ, Kennedy L: Effect of in vitro nonenzymatic glycosylation of human skin collagen on susceptibility to collagenase digestion. *Eur J Clin Invest* 1985;15:128–131.

123. Tanzer ML: Cross-linking of collagen: Endogenous aldehydes react in several ways to form a variety of unique covalent cross-links. *Science* 1973;180:561–566.

124. Tanzer ML: Isolation of lysinonorleucine from collagen. *Biochem Biophys Res Commun* 1970;39:183–189.

125. Heathcote JG, Bailey AJ, Grant ME: Studies on the assembly of the rat lens capsule. Biosynthesis of a cross-linked collagenous component of high molecular weight. *Biochem J* 1980;190:229–237.

126. Tanzer ML, Kefalides NA: Collagen cross-links: Occurrence in basement membrane collagens. *Biochem Biophys Res Commun* 1973;51:775–780.

127. Wu V-Y, Cohen MP: Reducible cross-links in human glomerular basement membrane. *Biochem Biophys Res Commun* 1982;104:911–915.

128. Pinnell SR, Martin GR: The cross-linking of collagen and elastin: Enzymatic conversion of lysine and peptide linkage to δ-aminoadipic-δ-semialdehyde (allysine) by an extract from bone. *Proc Natl Acad Sci USA* 1968;61:708–716.

129. Siegel RC, Martin GR: Collagen cross-linking: Enzymatic synthesis of lysine-derived aldehydes and the production of cross-linked components. *J Biol Chem* 1970;245:1653–1658.

130. Bailey AJ, Robins SP, Tanner MJA: Reducible components in the proteins of human erythrocyte membrane. *Biochim Biophys Acta* 1976;434:51–57.

131. Tanzer ML, Fairweather R, Gallop PM: Collagen cross-links: Isolation of reducible N-hexosyl hydroxylysine from borohydride reduced calf skin insoluble collagen. *Arch Biochem Biophys* 1972;48:76–84.

132. Urdanivia E, Cohen MP: Identification of amino acid sites of [^{14}C]-glycosylation of basement membranes. *Clin Res* 1981;29:734A.

133. Guitton, J-D., LePape A, Muh J-P: Influence of in vitro nonenzymatic glycosylation on the physico-chemical parameters of Type I collagen. *Coll Relat Res* 1984;4:253–264.

134. Guitton J-D, LePape A, Sizaret PY, Muh, J-P: Effect of in vitro glucosylation of Type I collagen fibrillogenesis. *Biosci Rep* 1981;1:945–954.

135. LePape A, Guitton J-D, Muh J-P: Distribution of nonenzymatically bound glucose in in vivo and in vitro glycosylated Type I collagen molecules. *FEBS Lett* 1984;170:23–27.

136. Lien Y-H, Stern R, Fu JCC, et al: Inhibition of collagen fibril formation in vitro and subsequent cross-linking of glucose. *Science* 1984;225:1489–1491.

137. Li W, Shen S, Robertson GA, et al: Increased solubility of newly synthesized collagen in retinal capillary pericyte cultures by nonenzymatic glycosylation. *Ophthalmic Res* 1984;16:315–321.

138. Yue DK, McLennan S, Delbridge L, et al: The thermal stability of collagen in diabetic rats: Correlation with severity of diabetes and nonenzymatic glycosylation. *Diabetologia* 1983;24:282–285.

139. Kohn RR, Cerami A, Monnier VM: Collagen aging in vitro by non-enzymatic glycosylation and browning. *Diabetes* 1984;33:57–59.

140. Yosha SF, Elden HR, Rabinovitch A, et al: Experimental diabetes mellitus and age-stimulated changes in intact rat dermal collagen. *Diabetes* 1983;32:739–742.

141. Brownlee M, Pongor S, Cerami A: Covalent attachment of soluble proteins by nonenzymatically glycosylated collagen: Role in the in situ formation of immune complexes. *J Exp Med* 1983;158:1739–1744.

142. Brownlee M, Vlassara H, Cerami A: Nonenzymatic glycosylation products on collagen covalently trap low-density lipoprotein. *Diabetes* 1985;34:938–941.

143. Michael AF, Brown DM: Increased concentration of albumin in kidney basement membranes in diabetes mellitus. *Diabetes* 1981;30:843–846.

144. Kato Y, Matsuda T, Watanabe K, et al: Alteration of ovalbumin immunogenic activity by glycosylation through Maillard reaction. *Agric Biol Chem* 1985;49:423–427.

145. Bassiouny AR, Rosenberg H, McDonald TL: Glucosylated collagen is antigenic. *Diabetes* 1983;32:1182–1184.

146. Cohen RA, Mauer SM, Barbosa J: Immunofluorescence studies of

skeletal muscle extracellular membranes in diabetes mellitus. *Lab Invest* 1978;29:275–278.

147. Chavers B, Etzwiler D, Michael AF: Albumin deposition in dermal capillary basement membranes in insulin-dependent diabetes mellitus. *Diabetes* 1981;30:275–278.

148. Ruoslahti E, Engvall E, Hagman EG: Fibronectin: Current concepts of its structure and function. *Collagen Res* 1981;1:95–128.

149. Yamada KM: Cell surface interactions with extracellular materials. *Annu Rev Biochem* 1983;52:761–799.

150. Mosher DF: Physiology of fibronectin. *Annu Rev Med* 1984;35:561–565.

151. Linder E, Vaheri A, Ruoslahti E, et al: Distribution of fibronectin in human tissues and relationships to other connective tissue components. *Ann NY Acad Sci* 1978;312:151–159.

152. Stenman S, Vaheri A: Distribution of a major connective tissue protein, fibronectin, in normal human tissues. *J Exp Med* 1978;147:1054–1064.

153. Laurie GW, LeBlond CP, Martin GR: Light microscopic immunolocalization of Type IV collagen, laminin, heparan sulfate proteoglycan and fibronectin in the basement membranes of a variety of rat organs. *Am J Anat* 1983;157:71–82.

154. Mosher DF, Sakesla O, Keski-Oja J, et al: Distribution of a major surface-associated glycoprotein, fibronectin, in cultures of adherent cells. *J Supramol Struct* 1977;6:551–557.

155. Oberly TD, Mosher DF, Mills MD: Localization of fibronectin within the renal glomerulus and its production by cultured glomerular cells. *Am J Pathol* 1979;96:651–662.

156. Courtoy PJ, Kanwar YS, Hynes RO, et al: Fibronectin localization in the rat glomerulus. *J Cell Biol* 1980;87:691–696.

157. Martinez-Hernandez A, Marsh CA, Clark CC, et al: Fibronectin: Its relationship to basement membranes. II. Ultrastructural studies. *Collagen Res* 1981;5:405–418.

158. Oberly TD, Murphy-Ullrich JE, Albrecht RM, et al: The effect of the dimeric and multimeric forms of fibronectin on the adhesion and growth of primary glomerular cells. *Exp Cell Res* 1983;145:265–276.

159. Cohen MP, Ku L: Inhibition of fibronectin binding to matrix components by nonenzymatic glycosylation. *Diabetes* 1984;33:970–974.

160. Tarsio JF, Wigness B, Rhode TD, et al: Nonenzymatic glycation of fibronectin and alterations in the molecular association of cell matrix and basement membrane components in diabetes mellitus. *Diabetes* 1985; 34:477–484.

161. Yamada KM, Kennedy DW, Kimata K, et al: Characterization of fibronectin interactions with glycosaminoglycans and identification of active proteolytic fragments. *J Biol Chem* 1980;255:6055–6063.

162. McKeown-Longo PJ, Mosher DF: Binding of plasma fibronectin to cell layers of human skin fibroblasts. *J Cell Biol* 1983;97:466–472.

163. Monnier VM, Cerami A: The search for non-enzymatic browning products in the human lens. *Invest Ophthalmol Vis Sci* 1981;20:169.

164. Monnier VM, Cerami A: Nonenzymatic browning in vivo: Possible process for aging for long-lived proteins. *Science* 1981;211:491–493.

165. Pirie A: Color and solubility of the proteins of human cataracts. *Invest Opthalmol* 1968;7:634-650.
166. DeBats A, Rhodes EL: Innate fluorescence in diabetic and aged kidneys. *Lancet* 1974;1:137-138.
167. Reynolds TM: Chemistry of nonenzymatic browning. *Adv Food Res* 1963; 12:1-52.
168. Reynolds TM: Chemistry of nonenzymatic browning. *Adv Food Res* 1965; 14:167-283.
169. Brownlee M, Vlassara H, Cerami A: Nonenzymatic glycosylation and the pathogenesis of diabetic complications. *Ann Intern Med* 1984;101:527-537.
170. Stevens VJ, Monnier VM, Cerami A: Hemoglobin glycosylation as a model for modification of other proteins. *Texas Rep Biol Med* 1980-1981; 40:387-396.
171. Monnier VM, Cerami A: Nonenzymatic glycosylation and browning of proteins in diabetes. *Clin Endocrinol Metab* 1982;11:431-452.
172. Monnier VM, Cerami A: Detection of nonenzymatic browning products in the human lens. *Biochim Biophys Acta* 1983;760:97-103.
173. Kent MJ, Light ND, Bailey AJ: Evidence for glucose-mediated covalent cross-linking of collagen after glycosylation in vitro. *Biochem J* 1985; 225:745-752.
174. Monnier VM, Kohn RR, Cerami A: Accelerated age-related browning of human collagen in diabetes mellitus. *Proc Natl Acad Sci USA* 1984;81:583-587.
175. Pongor S, Ulrich PC, Bencsath A, et al: Aging of proteins: Isolation and identification of a fluorescent chromophore from the reaction of polypeptides with glucose. *Proc Natl Acad Sci USA* 1984;81:2684-2688.
176. Chang JCF, Ulrich PC, Bucala R, et al: Detection of an advanced glycosylation product bound to protein in situ. *J Biol Chem* 1985; 260:7970-7974.
177. Brownlee M, Ulrich P, Vlassara H, et al: A furan-containing non-enzymatic glycosylation product causes P-450 enzyme-mediated cyto-toxicity. *Diabetes* 1985;34(suppl 1):51A.
178. Sakurai T, Takahashi H, Tsuchiya S: New Fluorescence of nonenzyma-tically glucosylated human serum albumin. *FEBS Lett* 1984;176:27-31.
179. Eble AS, Thorpe S, Baynes JW: Nonenzymatical glucosylation and glucose-dependent cross-linking of protein. *J Biol Chem* 1983;258:9406-9412.
180. Rucklidge GJ, Bates GP, Robins SP: Preparation and analysis of the products of non-enzymatic glycosylation and their relationship to cross-linking of proteins. *Biochim Biophys Acta* 1983;747:165-170.
181. Vlassara H, Brownlee M, Cerami A: High-affinity-receptor-mediated uptake and degradation of glucose-modified proteins: A potential mechanism for removal of senescent macromolecules. *Proc Natl Acad Sci USA* 1985;82:5588-5592.

Bibliography

Abraham EC, Huff TA, Cope ND, et al: Determination of the glycosylated hemoglobin (HbA$_{1c}$) with a new microcolumn procedure. *Diabetes* 1978; 27:931–937.

Abraham EC, Perry RE, Stallings M: Application of affinity chromatography for separation and quantitation of glycosylated hemoglobin. *J Lab Clin Med* 1983;102:187–197.

Achord DT, Brot FE, Bell CE, et al: Human β-glucuronidase: In vivo clearance and in vitro uptake by a glycoprotein recognition system on reticuloendothelial cells. *Cell* 1978;15:269–278.

Achord DT, Brot FE, Sly WS: Inhibition of the rat clearance system for agalacto-orosomucoid by yeast mannans and by mannose. *Biochem Biophys Res Commun* 1977;77:409–415.

Adhikari HR, Tappel AL: Fluorescent products in a glucose-glycine browning reaction. *J Food Sci* 1973;38:486–488.

Agardh C-D, Tallroth G: Lack of correlation between glycosylated haemoglobin concentrations and number of daily insulin injections: Cross sectional study in care of ambulatory diabetes. *Br Med J* 1985;291:622.

Aleyassine H: Low proportions of glycosylated hemoglobin associated with hemoglobin S and hemoglobin C. *Clin Chem* 1979;25:1484–1486.

Aleyassine H: Glycosylation of hemoglobin S and hemoglobin C. *Clin Chem* 1980;26:526–527.

Aleyassine H, Gardiner RJ, Blankstein LA, et al: Agar gel electrophoresis determination of glycosylated hemoglobin: Effect of variant hemoglobins, hyperlipidemia and temperature. *Clin Chem* 1981;27:472–475.

Aleyassine H, Gardiner RJ, Tonks DB, et al: Glycosylated hemoglobin in diabetes mellitus: Correlations with fasting plasma glucose, serum lipids and glycosuria. *Diabetes Care* 1980;3:508,514.

Allen, DW, Schroeder, WA, Balog J: Observation on the chromatographic heterogeneity of normal adult and fetal hemoglobin: A study of the effect of crystallization and chromatography on the heterogeneity and isoleucine content. *J Am Chem Soc* 1958;80:1628–1634.

Andressan TT, Seyer-Hansen K, Bailey AJ: Thermal stability, mechanical properties and reducible crosslinks of rat-tail tendon in experimental diabetes. *Biochim Biophys Acta* 1981;677:313–317.

Ansari, NH, Awasthi, YC, Srivastava, SK: Presented at the Annual Spring Meeting of the Association for Research in Vision and Ophthalmology, Florida, 1979.

Ansari NHM, Awasthi YL, Srivastava SK: Role of glycosylation in protein disulfide formation and cataractogenesis. *Exp Eye Res* 1980;31:9–19.

Arnqvist H, Cederblad G, Hermansson G, et al: A chromatographic method for measuring haemoglobin A_1: Comparison with two commercial kits. *Ann Clin Biochem* 1981;18:240–242.

Arturson G, Garby L, Robert M, et al: Oxygen affinity of whole blood in vivo and under standard conditions in control subjects with diabetes mellitus. *Scand J Clin Lab Invest* 1974;34:19–22.

Ashwell G, Morell AG: The role of surface carbohydrates in the hepatic recognition and transport of circulating glycoproteins. *Adv Enzymol* 1974; 41:99–128.

Bailey AJ, Robins SP, Tanner MJA: Reducible components in the proteins of human erythrocyte membrane. *Biochim Biophys Acta* 1976;434:51–57.

Banerjee RN, Sahni AL, Kumar V, et al: Antithrombin 3 deficiency in maturity onset diabetes mellitus and atherosclerosis. *Thromb Diath Haemorrh* 1974; 31:339–345.

Baynes JW, Thorpe SR, Murtiashaw MH: Nonenzymatic glucosylation of lysine residues in albumin. *Methods Enzymol* 1984;106:88–98.

Baxi L, Barad D, Reece EA, et al: Use of glycosylated hemoglobin as a screen for macrosomia in gestational diabetes. *Obstet Gynecol* 1984;64:347–350.

Bassiouny AR, Rosenberg H, McDonald TL: Glucosylated collagen is antigenic. *Diabetes* 1983;32:1182–1184.

Bernstein RE: Glycosylated hemoglobins: Hematological considerations determine which assay for glycohemoglobin is advisable. *Clin Chem* 1980; 26:174–175.

Bernstein RE: Glycosylated haemoglobin in glucose-6-phosphate dehydrogenase deficiency, letter to the editor. *Lancet* 1985;2:332–333.

Beswick HT, Harding JJ: Conformational changes induced in bovine lens α-crystallin by carbamylation. Relevance to cataract *Biochem J* 1984;223:221–227.

Blanc MH, Barnett DM, Gleason RE, et al: Hemoglobin A_{1c} compared with three conventional measures of diabetes control. *Diabetes Care* 1981;4:349–353.

Boden G, Master RW, Gordon SS, et al: Monitoring metabolic control in diabetic outpatients with glycosylated hemoglobin. *Ann Intern Med* 1980; 92:357–360.

Bolli G, Compagnucci P, Cartechini MG, et al: HBA$_1$ in subjects with abnormal glucose tolerance but normal fasting plasma glucose. *Diabetes* 1980;29:272–277.

Bolli G, Compagnucci P, Cartechini MC, et al: Analysis of short-term changes in reversibly and irreversibly glycosylated HBA$_1$: Relevance to diabetes mellitus. *Diabetologia* 1981;21:70–72.

Bookchin RM, Gallop PM: Structure of hemoglobin A$_{1c}$: Nature of the N-terminal β-chain blocking group. *Biochem Biophys Res Commun* 1968;32:86–93.

Botterman P: Rapid fluctuations in glycosylated haemoglobin concentration. *Diabetologia* 1981;20:159.

Boucher BJ, Welch SG, Beer MS: Glycosylated haemoglobins in the diagnosis of diabetes mellitus and for the assessment of chronic hyperglycemia. *Diabetologia* 1981;21:34–36.

Bouriotis V, Stott J, Galloway A, et al: Measurement of glycosylated hemoglobins using affinity chromatography. *Diabetologia* 1981;21:579–580.

Brooks AP, Nairn IM, Baird JD: Changes in glycosylated hemoglobin after poor control in insulin dependent diabetes. *Br Med J* 1980;281:707–710.

Brownlee M, Pongor S, Cerami A: Covalent attachment of soluble proteins by nonenzymatically glycosylated collagen: Role in the in situ formation of immune complexes. *J Exp Med* 1983;158:1739–1744.

Brownlee M, Ulrich P, Vlassara H, et al: A furan-containing nonenzymatic glycosylation product causes P-450 enzyme-medicated cytotoxicity. *Diabetes* 1985;34(suppl 1):51A.

Brownlee M, Vlassara H, Cerami A: Measurement of glycosylated amino acids and peptides from urine of diabetic patients using affinity chromatography. *Diabetes* 1980;29:1044–1047.

Brownlee M, Vlassara H, Cerami A: Nonenzymatic glycosylation reduces the susceptibility of fibrin to degradation by plasmin. *Diabetes* 1983;32:680–684.

Brownlee M, Vlassara H, Cerami A: Nonenzymatic glycosylation and the pathogenesis of diabetic complications. *Ann Intern Med* 1984;101:527–537.

Brownlee M, Vlassara H, Cerami A: Inhibition of heparin-catalyzed human antithrombin III activity by nonenzymatic glycosylation. *Diabetes* 1984; 33:532–535.

Brownlee M, Vlassara H, Cerami A: Nonenzymatic glycosylation products on collagen covalently trap low-density lipoproteins. *Diabetes* 1985;34:938–941.

Buckingham BA, Uitto J, Sandberg C, et al: Scleroderma-like changes in insulin-dependent diabetes mellitus: Clinical and biochemical studies. *Diabetes Care* 1984;7:163–169.

Bunn HF: Nonenzymatic glycosylation of protein: Relevance in diabetes. *Am J Med* 1981;70:325–330.

Bunn HF: Non-enzymatic glycosylation of protein: A form of molecular aging. *Schweiz Med Wochenschr* 1981;111:1503–1507.

Bunn HF, Briehl RW: The interaction of 2,3-diphosphoglycerate with various human hemoglobins. *J Clin Invest* 1970;49:1088–1095.

Bunn HF, Gabbay KH, Gallop PM: The glycosylation of hemoglobin: Relevance to diabetes mellitus. *Science* 1978;200:21–27.

Bunn HF, Haney DN, Gabbay KH, et al: Further identification of the nature and linkage of the carbohydrate in hemoglobin A_{1c}. *Biochem Biophys Res Commun* 1975;67:103–109.

Bunn HF, Haney DN, Kamin S, et al: The biosynthesis of human hemoglobin A_{1c}: Slow glycosylation of hemoglobin in vivo. *J Clin Invest* 1976;57:1652–1659.

Bunn HF, Higgins PJ: Reaction of monosaccharides with proteins: Possible evolutionary significance. *Science* 1981;213:222–224.

Bunn HF, Shapiro R, McManus M, et al: Structural heterogeneity of human hemoglobin A due to nonenzymatic glycosylation. *J Biol Chem* 1979; 254:3892–3898.

Calvert GD, Graham JJ, Mannick T, et al: Effects of therapy on plasma high-density lipoprotein-cholesterol concentration in diabetes mellitus. *Lancet* 1978;2:66–68.

Cammer W, Brosnan CF, Bloom BR, et al: Degradation of P_0, P_1, and Pr proteins in peripheral nervous system myelin by plasmin: Implications regarding the role of macrophages in demyelinating diseases. *J Neurochem* 1981;36:1506–1514.

Cederholm J, Ronquist G, Wibell L: Comparison of glycosylated hemoglobin with oral glucose tolerance test. *Diabete Metab* 1984;10:224–229.

Cerami A, Stevens VJ, Monnier VM: Role of nonenzymatic glycosylation in the development of the sequelae of diabetes mellitus. *Metabolism* 1979; 28:431–437.

Ceriello A, DelloRusso P, Curcio P, et al: Acetylsalicylic acid and lysine inhibit protein glycosylation in vitro: A preliminary report. *Diabete Metab* 1984;10:128–129.

Chang AY, Noble RE: 5-Hydroxymethylfurfural-forming proteins in the renal glomeruli of control and streptozotocin-diabetic rats. *Life Sci* 1980;26:1329–1333.

Chang JCF, Ulrich PC, Bucala R, et al: Detection of an advanced glycosylation product bound to protein in situ. *J Biol Chem* 1985;260:7970–7974.

Chase HP, Glasgow AM: Juvenile diabetes mellitus and serum lipids and lipoprotein levels. *Am J Dis Child* 1976;130:1113–1117.

Chavers B, Etzwiler D, Michael AF: Albumin deposition in dermal capillary basement membranes in insulin-dependent diabetes mellitus. *Diabetes* 1981; 30:275–278.

Chiou SH, Chylack LT, Bunn HF, et al: Role of nonenzymatic glycosylation in experimental cataract formation. *Biochem Biophys Res Commun* 1980; 95:894–901.

Chiou SH, Chylack LT, Tung WH: Nonenzymatic glycosylation of bovine lens crystallins: Effect of aging. *J Biol Chem* 1981;256:5176–5180.

Chou J, Robinson CA Jr, Siegel AL: Simple method for estimating glycosylated hemoglobins and its application to evaluation of diabetic patients. *Clin Chem* 1978;24:1708–1710.

Citrin W, Ellis GJ, Skyler JS: Glycosylated hemoglobin: A tool in identifying psychological problems. *Diabetes Care* 1980;3:563–564.

Clarke JT, Canivet J: Hemoglobin A_{1c} separation by microcolumn chromatography: A new rapid method of assay. *Diabete Metab* 1979;5:293–296.

Clegg MD, Schroeder WA: A chromatographic study of the minor components of normal adult human hemoglobin including a comparison of hemoglobin from normal and phenylketonuric individuals. *J Am Chem Soc* 1959; 81:6065–6069.

Cohen MP, Ku L: Inhibition of fibronectin binding to matrix components by nonenzymatic glycosylation. *Diabetes* 1984;33:970–974.

Cohen MP, Surma ML: [^{35}S]-Sulfate incorporation into glomerular basement membrane glycosaminoglycans is decreased in experimental diabetes. *J Lab Clin Med* 1981;98:715–722.

Cohen MP, Surma ML: Effect of diabetes on in vivo metabolism of [^{35}S]-labeled glomerular basement membrane. *Diabetes* 1984;33:8–12.

Cohen MP, Urdanivia E, Surma M, et al: Nonenzymatic glycosylation of basement membranes. In vitro studies. *Diabetes* 1981;30:367–371.

Cohen MP, Urdanivia E, Surma M, et al: Increased glycosylation of glomerular basement membrane collagen in diabetes. *Biochem Biophys Res Commun* 1980;95:765–769.

Cohen MP, Urdanivia E, Wu V-Y: Nonenzymatic glycosylation of basement membrane *Renal Physiol* 1981;4:90–95.

Cohen MP, Wu, V-Y: Identification of specific amino acids in diabetic glomerular basement membrane collagen subject to nonenzymatic glucosylatio in vivo. *Biochem Biophys Res Commun* 1981;100:1549–1554.

Cohen MP, Wu V-Y: Age-related changes in nonenzymatic glycosylation of human basement membranes. *Exp Gerontol* 1983;18:461–469.

Cohen RA, Mauer SM, Barbosa J, et al: Immunofluorescence studies of skeletal muscle extracellular membranes in diabetes mellitus. *Lab Invest* 1978;29:275–278.

Cohenford MA, Urbanowski JC, Shepard DC, et al: Nonenzymatic glycosylation of human IgG: *In vitro* preparation. *Immunol Comm* 1983;12:189–200.

Cole RA: A new test for diabetes mellitus: The measurement of hemoglobin A_{1c} and the total fast hemoglobin using high pressure liquid chromatography. *Lab Management* 1978;16:41–44.

Cole RA, Soeldner JS, Dunn PH, et al: A rapid method for the determination of glycosylated hemoglobins using high pressure liquid chromatography. *Metabolism* 1978;27:289–301.

Coller BS, Frank RN, Milton RC, et al: Plasma cofactors of platelet function: Correlation with diabetic retinopathy and hemoglobin A_{Ia-c}. *Ann Intern Med* 1978;88:311–316.

Coradello H, Pollak A, Pagano M, et al: Nonenzymatic glycosylation of Cathepsin B.: Possible influence on conversion of proinsulin to insulin. *IRCS Med Sci* 1981;9:766–767.

Cotlier E: Aspirin effect on cataract formation in patients with rheumatoid arthritis alone or combined with diabetes. *Int Ophthalmol* 1981;3:173–177.

Courtoy PJ, Kanwar YS, Hynes RO, et al: Fibronectin localization in the rat glomerulus. *J Cell Biol* 1980;87:691–696.

Curtiss LK, Witztum JL: A novel method for generating region-specific monoclonal antibodies to modified proteins. Application to the identification of human glucosylated low-density lipoproteins. *J Clin Invest* 1983; 72:1427–1438.

Curtiss LK, Witztum JL: Plasma apolipoproteins Al, AlI, B, Cl and E are glucosylated in hyperglycemic diabetic subjects. *Diabetes* 1985;34:452–461.

Dahlquist G, Blom L, Bolme P, et al: Metabolic control in 131 juvenile-onset diabetic patients as measured by HbA_{1C}: Relation to age, duration, C-peptide, insulin dose, and one or two insulin injections. *Diabetes Care* 1982; 5:399–403.

Daubresse JC, Lemy C, Bailly A, et al: The usefulness of a rapid method for total fast hemoglobins determination in screening for diabetes control. *Diabete Metab* 1979;5:301–305.

David JE, McDonald JM, Jarret L: A high performance liquid chromatography method for hemoglobin A_{1c}. *Diabetes* 1978;27:102–107.

Day JF, Ingelbretsen CG, Ingelbretsen WR, et al: Nonenzymatic glucosylation of serum proteins and hemoglobin: Response to changes in blood glucose levels in diabetic rats. *Diabetes* 1980;29:524–527.

Day JF, Thornburg RW, Thorpe SW, et al: Nonenzymatic glucosylation of rat albumin: Studies in vitro and in vivo. *J Biol Chem* 1979;254:9393–9400.

Day JF, Thorpe SR, Baynes JW: Non-enzymatically glucosylated albumin: In vitro preparation and isolation from normal human serum. *J Biol Chem* 1979;254:595–597.

DeBats A, Rhodes EL: Innate fluorescence in diabetic and aged kidneys. *Lancet* 1974;1:137–138.

deBoer M-J, Miedema K, Casparie AF: Glycosylated haemoglobin in renal failure. *Diabetologia* 1980;18:437–440.

Ditzel J: Changes in red cell oxygen release capacity in diabetes mellitus. *Fed Proc* 1977;38:2484–2488.

Ditzel J: Affinity hypoxia as a pathogenetic factor of microangiopathy with particular reference to diabetic retinopathy. *Acta Endocrinol* 1980;94:39–55.

Ditzel J, Anderson H, Peters ND: Oxygen affinity of hemoglobin and red cell 2,3-diphosphoglycerate in childhood diabetes. *Acta Pediatr Scand* 1975; 64:355–361.

Ditzel J, Daugaard P, Anderson H: Oxygen affinity of haemoglobin and red cell 2,3-diphosphoglycerate in childhood diabetes. *Diabetologia* 1974; 10:363.

Ditzel J, Jaeger P, Standl E: An adverse effect of insulin on the oxygen release capacity of red blood cells in nonacidotic diabetes. *Metabolism* 1978;27:929–934.

Ditzel J, Kawahara R, Mourtis-Andersen T, et al: Changes in blood glucose, glycosylated hemoglobin and hemoglobin-oxygen affinity following meals in diabetic children. *Eur J Pediatr* 1981;137:171–174.

Ditzel J, Kjaegaard J-J, Kawahara R, et al: Glycosylated hemoglobin in relation to rapid fluctuations in blood glucose in children with insulin dependent diabetes: A comparison of methods with and without prior dialysis. *Diabetes Care* 1981;4:551–555.

Ditzel J, Nielsen NV, Kjaergaard JJ: Hemoglobin A and red cell oxygen release capacity in relation to early retinal changes in newly discovered overt and chemical diabetes. *Metabolism* 1979;28(suppl 1):440–447.

Ditzel J, Standl E: The problem of tissue oxygenation in diabetes mellitus. *Acta Med Scand* 1978; suppl. 578, pp59–68.

Dix D, Cohen P, Kingsley S, et al: Glycohemoglobin and glucose tolerance tests compared as indicators of borderline diabetes. *Clin Chem* 1979;25:877–879.

Dods RF, Bolmey C: Glycosylated hemoglobin assay and oral glucose tolerance test compared for detection of diabetes mellitus. *Clin Chem* 1979; 25:764–768.

Dolhofer R, Renner R, Wieland OH: Different behavior of haemoglobin A_{1a-c} and glycosyl-albumin levels during recovery from diabetic ketoacidosis and non-acidotic coma. *Diabetologia* 1981;21:211–215.

Dolhofer R, Siess EA, Wieland OH: Inactivation of bovine kidney β-N-acetyl-D-glucosaminidase by nonenzymatic glucosylation. *Hoppe Seylers Z Physiol Chem* 1982;363:1427–1436.

Dolhofer R, Wieland OH: In vitro glycosylation of hemoglobin by different sugars and sugar phosphates. *FEBS Lett* 1978;85:86–90.

Dolhofer R, Wieland OH: Glycosylation of serum albumin: Elevated glycosyl-albumin in diabetic patients. *FEBS Lett* 1979;103:282–286.

Dolhofer R, Wieland OH: Preparation and biological properties of glycosylated insulin. *FEBS Lett* 1979;100:133–136.

Dolhofer R, Wieland OH: Increased glycosylation of serum albumin in diabetes mellitus. *Diabetes* 1980;29:417–422.

Donde UM, Baxi AJ, El Tawil H, et al: Glycosylated haemoglobin in glucose-6-phosphate dehydrogenase deficiency, letter to the editor. *Lancet* 1985; 2:47.

Dunn PJ, Cole RA, Soeldner JS, et al: Reproducibility of hemoglobin A and sensitivity to various degrees of glucose intolerance. *Ann Intern Med* 1979; 91:390–396.

Dunn PJ, Cole RA, Soeldner JS, et al: Stability of hemoglobin A_{1c} levels on repetitive determination in diabetic outpatients. *J Clin Endocrinols Metab* 1981;52:1019– 1022.

Dunn PJ, Cole RA, Soeldner JS, et al: Temporal relationship of glycosylated hemoglobin concentrations to glucose control in diabetes. *Diabetologia* 1979; 17:213–220.

Eble AS, Thorpe S, Baynes JW: Nonenzymatic glucosylation and glucose-dependent cross-linking of protein. *J Biol Chem* 1983;258:9406–9412.

Elkeles RJ, Wu J, Hambley J: Hemoglobin A_1 blood glucose and high density lipoprotein cholesterol in insulin requiring diabetes. *Lancet* 1978;2:547–548.

Ellis G, Diamandis EP, Giesbrecht EE, et al: An automated high pressure liquid chromatographic assay for hemoglobin A. *Clin Chem* 1984;30:1746–1752.

Eross J, Kreutzmann D, Jimenez M, et al: Colorimetric measurement of glycosylated protein in whole blood, red blood cells, plasma, and dried blood. *Ann Clin Biochem* 1984;21:477–483.

Esnard F, Guitton JD, Stauber WT, et al: Nonenzymatic glycosylation of rat serum proteinase inhibitors and change in their concentration during experimental diabetes. *Molecular Physiology* 1985;7:211–218.

Fadel HE, Hammond SD, Huff TA, et al: Glycosylated hemoglobins in normal pregnancy and gestational diabetes mellitus. *Obstet Gynecol* 1979;54:322–326.

Farris L, Wajcman H, Jones RT, et al: Functional properties of hemoglobin A_{1C}. *Clin Res* 1977;25:115A.

Finot P-A: Nonenzymatic browning products: Physiologic effects and metabolic transit in relation to chemical structure. A review. *Diabetes* 1982;31(suppl 3):22–28.

Finot PA, Bricout J, Viani R, et al: Identification of a new lysine derivative obtained upon acid hydrolysis of heated milk. *Experientia* 1968;24:1097–1099.

Finot P-A, Bujard E, Mottu F, et al: Availability of the true Schiff bases of lysine: Chemical evaluation of the Schiff base between lysine and lactose in milk. *Adv Exp Med Biol* 1977;86:343–365.

Finot PA, Viani R, Bricout J, et al: Detection and identification of pyridosine, a second lysine derivative obtained upon acid hydrolysis of heated milk. *Experientia* 1969;25:134–135.

Fischer W, deJong C, Voigt E, et al: The colorimeteric determination of HbA in normal and diabetic subjects. *Clin Lab Haematol* 1980;2:129–138.

Fisher RW, Winterhalter KH: The carbohydrate moiety in hemoglobin A_{1C} is present in the ring form. *FEBS Lett* 1981;135:145–147.

Fitzgerald MD, Cauchi MN: Glycosylated hemoglobins in patients with a hemoglobinopathy. *Clin Chem* 1980;26:360–361.

Fitzgibbons JF, Koler RD, Jones RT: Red cell age-related changes of hemoglobin A_{1a+b} and A_{1C} in normal and diabetic subjects. *J Clin Invest* 1976;58:820–824.

Flock EV, Bennett PH, Savage PJ, et al: Bimodality of glycosylated hemoglobin distribution in Pima Indians. *Diabetes* 1979;28:984–989.

Freinkel N: Of pregnancy and progeny. *Diabetes* 1980;29:1023–1035.

Flückiger R, Gallop P: Measurement of nonenzyumatic protein glycosylation. *Methods Enzymol* 1984;106:77–87.

Flückiger R, Harmon W, Meier W, et al: Hemoglobin carbamylation in uremia. *N Engl J Med* 1981;304:823–827.

Flückiger R, Winterhalter KH: In vitro synthesis of hemoglobin A_{1C}. *FEBS Lett* 1976;71:356–360.

Gabbay KH, Hasty K, Breslow JL, et al: Glycosylated hemoglobin and long term blood glucose control in diabetes mellitus. *J Clin Endocrinol Metab* 1977;44:859–864.

Gabbay KH, Sosenko JM, Banuchi GA, et al: Increased glycosylation of hemoglobin A in diabetic patients. *Diabetes* 1979;28:337–340.

Gallop PM, Flückiger R, Hanneken A, et al: Chemical quantitation of hemoglobin glycosylation: Fluorometric detection of formaldehyde released upon periodate oxidation of glycoglobin. *Anal Biochem* 1981;117:427-432.

Garlick RL, Mazer JS, Higgins PJ, et al: Characterization of glycosylated hemoglobin: Relevance to monitoring of diabetic control and analysis of other proteins. *J Clin Invest* 1983;71:1062-1072.

Garrick LM, McDonald MJ, Shapiro R, et al: Structural analysis of the minor human hemoglobin components: HbA_{1a_1}, HbA_{1a_2} and HbA_{1b}. *Eur J Biochem* 1980;106:353-359.

Ghiggeri GM, Candiano G, Delfino G, et al: Glycosyl albumin and diabetic microalbuminuria: Demonstration of altered renal handling. *Kidney Int* 1984;25:565-570.

Gillery P, Maquart F-X, Corcy J-M, et al: A glucose transfer from membrane glycoconjugates to hemoglobin in isolated young red blood cells: Another biosynthetic way for glycosylated haemoglobins. *Eur J Clin Invest* 1984;14:317-322.

Gillery P, Maquart F-X, Gattegno L, et al: A glucose-containing fraction extracted from young erythrocyte membrane is capable of transferring glucose to hemoglobin in vitro. *Diabetes* 1982;31:371-374.

Gillmer MDG, Beard RW, Brooks FW, et al: Carbohydrate metabolism in pregnancy: I Diurnal plasma glucose profile in normal and diabetic women. *Br Med J* 1975;3:399-404.

Goebel FD, Fuessel H, Dörfler H, et al: Short term changes of glycosylated haemoglobins during glucose administration in healthy and diabetic subjects. *Res Exp Med* 1981;179:1330-1334.

Goldstein DE, Walker B, Rawlings SS, et al: Hemoglobin A_{1C} levels in children and adolescents with diabetes mellitus. *Diabetes Care* 1980;3:503-507.

Goldstein JL, Ho YK, Basu SK, et al: Binding site on macrophages that mediates uptake and degradation of acetylated low density lipoprotein, producing massive cholesterol deposition. *Proc Natl Acad Sci USA* 1979;76:333-337.

Goldstein DE, Peth SB, England JD, et al: Effects of acute changes in blood glucose on HbA_{1C}. *Diabetes* 1980;29:623-628.

Gomo ZAR: The determination of glucose and glycosylated haemoglobin in a nondiabetic Zimbabwean African population. *Ann Clin Biochem* 1985;22:362-365.

Gonen B, Baenziger J, Schonfeld G, et al: Non-enzymatic glycosylation of low density lipoproteins in vitro: Effects of cell-interactive properties. *Diabetes* 1981;30:875-878.

Gonnen B, Rochman H, Rubenstein, AH: Metabolic control in diabetic patients: Assessment by hemoglobin A_1 values. *Metabolism* 1979 (suppl I):448-452.

Gonnen B, Rubenstein AH, Rochman H, et al: Hemoglobin A_1: An indicator of metabolic control of diabetic patients. *Lancet* 1977;2:734-737.

Graf RJ, Halter JB, Halar E, et al: Nerve conduction abnormalities in untreated maturity-onset diabetes: Relation to levels of fasting plasma glucose and glycosylated hemoglobin. *Ann Intern Med* 1979;90:298-303.

Graf RJ, Halter JB, Porte D: Glycosylated hemoglobin in normal subjects and subjects with maturity onset diabetes. *Diabetes* 1978;27:834–839.

Gragnoli G, Signorini AM, Tanganelli I: Nonenzymatic glycosylation of urinary proteins in Type I (Insulin-dependent) diabetes: Correlation with metabolic control and degree of proteinuria. *Diabetologia* 1984;26:411–414.

Gragnoli G, Tanganelli I, Signorini AM, et al: Nonenzymatic glycosylation of serum protein as an indicator of diabetic control. *Acta Diabetol Lat* 1982;19:161–166.

Gralnick HG, Coller BS, Sultan Y: Carbohydrate deficiency of the factor VIII/Von Willebrand factor protein in Von Willebrand's disease variants. *Science* 1976;192:56–59.

Gould BJ, Hall PM, Cook JGH: Measurement of glycosylated hemoglobin using an affinity chromatography method. *Clin Chem Acta* 1982;125:41–48.

Guitton J-D, LePape A, Muh J-P: Influence of in vitro nonenzymatic glycosylation on the physio-chemical parameters of Type I collagen. *Coll Rel Res* 1984;4:253–264.

Guitton J-D, LePape A, Sizaret PY, et al: Effect of in vitro nonenzymatic glucosylation on type I collagen fibrillogenesis. *Biosci Rep* 1981;1:945–954.

Guthrow CE, Morris MA, Day JF, et al: Enhanced nonenzymatic glucosylation of serum albumin in diabetes mellitus. *Proc Natl Acad Sci USA* 1979;76:4258–4261.

Hall, PM, Cook JGH, Sheldon J, et al: Glycosylated hemoglobins and glycosylated plasma proteins in the diagnosis of diabetes mellitus and impaired glucose tolerance. *Diabetes Care* 1984;7:147–150.

Hamlin CR, Kohn RR: Evidence for progressive, age-related structural changes in post-mature human collagen. *Biochim Biophys Acta* 1971;236:458–467.

Hamlin CR, Kohn RR, Luschin JH: Apparent accelerated aging of human collagen in diabetes mellitus. *Diabetes* 1975;24:902–904.

Hamman RF, Wells R, Ryschon K, et al: Glycohemoglobin stability. *Diabetes Care* 1982;5:143–144.

Haney DN, Bunn HF: Glycosylation of hemoglobin in vitro: Affinity labeling of hemoglobin by glucose-6-phosphate. *Proc Natl Acad Sci USA* 1976;73:3534–3538.

Hayashi Y, Makino M: Fluorometric measurement of glycosylated albumin in human serum. *Clin Chim Acta* 1985;149:13–19.

Heathcote JG, Bailey AJ, Grant ME: Studies on the assembly of the rat lens capsule: Biosynthesis of a cross-linked collagenous component of high molecular weight. *Biochem J* 1980;190:229–237.

Hekali R, Puuka R, Pokja R, et al: Abnormal Hb-variants can cause unexpected high values in the estimation of glycosylated hemoglobins. *J Clin Chem Clin Biochem* 1981;19:695–698.

Higgins PJ, Bunn HF: Kinetic analysis of the nonenzymatic glycosylation of hemoglobin. *J Biol Chem* 1981;256:5204–5208.

Higgins PJ, Garlick RL, Bunn HF: Glycosylated hemoglobin in human and animal red cells: Role of glucose permeability. *Diabetes* 1982;31:743–748.

Hill DE: Fetal effects of insulin. *Obstet Gynecol Ann* 1982;11:133–149.

Hodge JE: Chemistry of browning reactions in model systems. *J Agric Food Chem* 1953;1:928–943.

Holmquist WR, Schroeder WA: Properties and partial characterization of adult human hemoglobin A_{1C}. *Biochim Biophys Acta* 1964;82:639–641.

Holmquist WR, Schroeder WA: A new N-terminal blocking group involving a Schiff base in hemoglobin A_{1C}. *Biochemistry* 1966;5:2489–2503.

Housley TJ, Tanzer ML: The separation and amino acid analysis of collagen crosslinks on an extended basic ion-exchange column. *Anal Biochem* 1981;114:310–315.

Husiman W, Kuijken JPAA, Tan-Tjiong HL, et al: Unstable glycosylated hemoglobin in patients with diabetes mellitus. *Clin Chim Acta* 1982;118:303–309.

Javid J, Pettis PK, Koenig RJ, et al: Immunologic characterization and quantification of haemoglobin A_{1C}. *Br J Haematol* 1978;38:329–337.

Jeraj KP, Michael AF, Mauer SM, et al: Glucosylated and normal human or rat albumin do not bind to renal basement membranes of diabetic and control rats. *Diabetes* 1983;32:380–382.

Johnson WJ, Pizzo SV, Imber MJ, et al: Receptors for maleylated proteins regulate secretion of neutral proteases by murine macrophages. *Science* 1982;218:574–576.

Jones RL: Fibrinopeptide-A in diabetes mellitus: Relation to levels of blood glucose, fibrinogen disappearance, and hemodynamic changes. *Diabetes* 1985;34:836–843.

Jones MB, Koler RD, Jones RT: Microcolumn method for the determination of hemoglobin minor fractions A_{1a+b} and A_{1C}. *Hemoglobin* 1978;2:53–58.

Jones IR, Owens DR, Williams S, et al: Glycosylated serum albumin: An intermediate index of diabetic control. *Diabetes Care* 1983;6:501–503.

Jones RL, Peterson CM: Reduced fibrinogen survival in diabetes mellitus: A reversible phenomenon. *J Clin Invest* 1979;63:485–493.

Jovanovic L, Peterson CM: The clinical utility of glycosylated hemoglobin. *Am J Med* 1981;70:331–338.

Kasai K, Nakamura T, Kase N, et al: Increased glycosylation of proteins from cataractous lenses in diabetes. *Diabetologia* 1983;25:36–38.

Kato Y, Matsuda T, Watanabe K, et al: Alteration of ovalbumin immunogenic activity by glycosylation through Maillard reaction. *Agric Biol Chem* 1985;49:423–427.

Kemp SF, Creech RH, Horn TR: Glycosylated albumin and transferrin: Short term markers of blood control. *J Pediatr* 1984;105:394–398.

Kennedy L, Baynes JW: Non-enzymatic glycosylation and the chronic complications of diabetes. *Diabetologia* 1984;26:93–98.

Kennedy AL, Kandell TW, Merimee TJ: Serum protein-bound hexose in diabetes: The effect of glycemic control. *Diabetes* 1979;28:1006–1010.

Kennedy AL, Lappin TRJ, Lavery TD, et al: Relation of high density lipoprotein cholesterol concentration to type of diabetes and its control. *Br Med J* 1978;2:1191–1194.

Kennedy AL, Mehl TD, Merimee TJ: Nonenzymatically glycosylated serum protein: Spurious elevation due to free glucose in serum. *Diabetes* 1980;29:413–415.

Kesaniemi YA, Witztum JL, Steinbrecher UP: Receptor-mediated clearance of low density lipoprotein in man: New estimates using glucosylated low density lipoprotein. *Arteriosclerosis* 1982;2:441A.

Kesson CM, Young RE, Talwar D, et al: Glycosylated hemoglobin in the diagnosis of non-insulin dependent diabetes mellitus. *Diabetes Care* 1982;5:395–398.

Kent MJ, Light ND, Bailey AJ: Evidence for glucose-mediated covalent cross-linking of collagen after glycosylation in vitro. *Biochem J* 1985;225:745–752.

Kim HJ, Kurup IV: Decreased catabolism of glycosylated low density lipoprotein in diabetic rats. *Diabetes* 1981;30:47A.

Kim HJ, Kurup IV: Nonenzymatic glycosylation of human plasma low density lipoprotein: Evidence for in vitro and in vivo glycosylation. *Metabolism* 1982;31:348–353.

Kinoshita JH: Mechanisms initiating cataract formation. *Invest Ophthalmol* 1974;13:713–724.

Kjaergaard J-J, Ditzel J: Hemoglobin A_{1C} as in index of long-term blood glucose regulation in diabetic pregnancy. *Diabetes* 1979;28:694–696.

Klenk DC, Hermanson GT, Korhn RI, et al: Determination of glycosylated hemoglobin by affinity chromatography: Comparison with colorimetric and ion exchange methods, and effects of common interferences. *Clin Chem* 1982;28:2088–2094.

Klujber L, Malnar D, Kardos M, et al: Metabolic control, glycosylated hemoglobin, and high density lipoprotein cholesterol in diabetic children. *Eur J Pediatr* 1979;132:289–297.

Koenig RJ, Blobstein SH, Cerami A: Structure of carbohydrate and hemoglobin A. *J Biol Chem* 1977;252:2992–2997.

Koenig RJ, Cerami A: Synthesis of hemoglobin A_{1C} in normal and diabetic mice: Potential model of basement membrane thickening. *Proc Natl Acad Sci USA* 1975;72:3687–3691.

Koenig RJ, Cerami A: Hemoglobin A_{1C} and diabetes mellitus. *Annu Rev Med* 1980;31:29–34.

Koenig RJ, Peterson CM, Jones RL, et al: The correlation of glucose regulation and hemoglobin A_{1C} in diabetes mellitus. *New Engl J Med* 1976;295:417–420.

Koenig RJ, Peterson CM, Kilo C, et al: Hemoglobin A_{1C} as an indicator of the degree of glucose intolerance in diabetes. *Diabetes* 1976;25:230–232.

Kohn RR, Cerami A, Monnier WM: Collagen aging in vitro by nonenzymatic glycosylation and browning. *Diabetes* 1984;33:57–59.

Koivisto VA, Ekblom M, Icen A, et al: Abnormal hemoglobin variant: A source of error in chromatographic determination of hbA_1. *Diabetes Care* 1982;5:650–651.

Kornfeld S: The effects of structural modification in the biologic activity of human transferrin. *Biochemistry* 1968;1:945–954.

Kraemer FB, Chen Y-DI, Cheung RMC, et al: Are the binding and degradation of low density lipoprotein altered in Type 2 (non-insulin dependent) diabetes mellitus? *Diabetologia* 1981;23:28–33.

Krishnamoorthy R, Gacon G, Labie D: Isolation and characterization of hemoglobin A$_{1b}$. *FEBS Lett* 1977;77:99–101.

Lanoe R, Soria J, Thibult N, et al: Glycosylated hemoglobin concentrations and clinical test results in insulin-dependent diabetes. *Lancet* 1977;2:1156–1157.

Laurie GW, LeBlond CP, Martin GR: Light microscopic immunolocalization of Type IV collagen, laminin, heparan sulfate proteoglycan and fibronectin in the basement membranes of a variety of rat organs. *Am J Anat* 1983;157:71–82.

Lee JH, Shin DH, Lupovitch A, et al: Glycosylation of lens proteins in senile cataract and diabetes mellitus. *Biochem Biophys Res Commun* 1984;123:888–893.

LePape A, Guitton JD, Gutman N, et al: Nonenzymatic glycosylation of collagen in diabetes: Incidence on increased normal platelet aggregation. *Haemostasis* 1983;13:36–54.

LePape A, Gutman N, Guitton JD, et al: Nonenzymatic glycosylation increases platelet aggregating potency of collagen from placenta of diabetic human beings. *Biochem Biophys Res Commun* 1983;111:602–610.

LePape A, Guitton J-D, Muh JP: Modification of glomerular basement membrane cross-links in experimental diabetic rats. *Biochem Biophys Res Commun* 1981;100:1214–1221.

LePape A, Guitton J-D, Muh JP: Distribution of nonenzymatically bound glucose in in vivo and in vitro glycosylated Type I collagen molecules. *FEBS Lett* 1984;170:23–27.

LePape A, Muh JP, Bailey AJ: Characterization of *N*-glycosylated Type I collagen in streptozotocin-induced diabetes. *Biochem J* 1981;197:405–412.

Leslie RDG, Pyke DA, John PN, et al: Haemoglobin AI in diabetic pregnancy. *Lancet* 1978;2:958–959.

Lev-Ran A: Glycohemoglobin: Its use in the follow-up of diabetes and diagnosis of glucose intolerance. *Arch Intern Med* 1981;141:747–749.

Lev-Ran A, Vanderlaan WP: Glycohemoglobins and glucose intolerance. *JAMA* 1979;241:912–914.

Li W, Shen S, Robertson GA, et al: Increased solubility of newly synthesized collagen in retinal capillary pericyte cultures by nonenzymatic glycosylation. *Ophthalmic Res* 1984;16:315–321.

Liang JN, Chakrabarti B: Glycosylation-induced conformational change in α-crystallin of bovine lens. *Biophys J* 1981;33:138A.

Liang JN, Chylack T: Change in the protein tertiary structure with nonenzymatic glycosylation of calf α-crystallin. *Biochem Biophys Res Commun* 1984;123:899–906.

Lien Y-H, Stern R, Fu JCC, et al: Inhibition of collagen fibril formation in vitro and subsequent cross-linking by glucose. *Science* 1984;225:1489–1491.

Lind T, Cheyne GA: Effect of normal pregnancy upon the glycosylated hemoglobins. *Br J Obstet Gynecol* 1979;86:210–213.

Linder E, Vaheri A, Ruoslahti E, et al: Distribution of fibronectin in human tissues and relationships to other connective tissue components. *Ann NY Acad Sci* 1978;312:151–159.

Lopez-Virella MF, Sherer GK, Lees Am, et al: Surface binding, internalization and degradation by cultured human fibroblasts of low density lipoproteins isolated from Type I (insulin-dependent) diabetic patients: Changes with metabolic control. *Diabetologia* 1982;22:430–436.

Lopez-Virella MFL, Stone PG, Colwell JA: Serum high density lipoprotein in diabetic patients. *Diabetologia* 1977;13:285–291.

Lorenzi M, Cagliero E, Markey B, et al: Interaction of human endothelial cells with elevated glucose concentrations and native and glycosylated low density lipoproteins. *Diabetologia* 1984;26:218–222.

Lubec G, Legenstein E, Pollack A, et al: Glomerular basement membrane changes, HbA_{1c} and urinary excretion of acid glycosaminoglycans in children with diabetes mellitus. *Clin Chim Acta* 1980;103:45–49.

Lunetta M, Infantone E, Spanti D, et al: Glycosylated hemoglobin in uraemic patients with normal glucose tolerance or insulin dependent diabetes. *IRCS Med Sci* 1981;9:844–845.

Lutjens A, teVelde AA, v.d.Veen EA, et al: Glycosylation of human fibrinogen in vivo. *Diabetologia* 1985;28:87–89.

Lyons TJ, Kennedy L: Effect of in vitro nonenzymatic glycosylation of human skin collagen on susceptibility to collagenase digestion. *Eur J Clin Invest* 1985;15:128–131.

Lyons TJ, Kennedy L: Non-enzymatic glycosylation of skin collagen in patients with Type I (insulin-dependent) diabetes mellitus and limited joint mobility. *Diabetologia* 1985;28:2–5.

Ma A, Naughton MA, Cameron DP: Glycosylated plasma protein: A simple method for elimination of interference by glucose in its estimation. *Clin Chim Acta* 1981;115:111–117.

Madsen H, Ditzel J, Hansen P, et al: Hemoglobin A_{1c} determinations in diabetic pregnancy. *Diabetes Care* 1981;4:541–546.

Maillard LC: Action des acides aminés sur les sucres: Formation des melanoides par voie methodique. *C R Acad Sci* 1912;154:66–68.

Maillard LC: Synthése des matiéres humiques par action des acides aminés sur les sucres reducteurs. *Ann Chim* 1916;6:258–317.

Mandrel SS, Shin DH, Newman BL, et al: Glycosylation in vivo of human lens capsule (basement membrane) and diabetes mellitus. *Biochem Biophys Res Commun* 1983;117:51–56.

Malia AK, Hermanson GT, Krohn RI, et al: Preparation and use of a boronic acid affinity support for separation and quantitation of glycosylated hemoglobins. *Anal Lett* 1981;14:649–661.

Martinez-Hernandez A, Marsh CA, Clark CC, et al: Fibronectin: Its relationship to basement membranes. II. Ultrastructural studies. *Collagen Res* 1981;5:405–418.

Marynen P, Vanleuven F, Cassiman J-J, et al: Proteolysis at a lysine residue abolishes the receptor-recognition site of alpha 2-macroglobulin complexes. *FEBS Lett* 1982;137:241–22.

Mazze RS, Shamoon H, Pasmantier R, et al: Reliability of blood glucose monitoring by patients with diabetes mellitus. *Am J Med* 1984;77:211–217.

McDermott K, Cooks M, Peterson CM: Patient determined glycosylated hemoglobin measurements: An aid to patient education. *Diabetes Care* 1981;4:480–483.

McDonald MJ, Bleichman M, Bunn HF, et al: Functional properties of the glycosylated minor components of human adult hemoglobin. *J Biol Chem* 1979;254:702–707.

McDonald JM, Davis JE: Glycosylated hemoglobins and diabetes mellitus. *Hum Pathol* 1979;10:279–290.

McDonald MJ, Shapiro R, Bleichman M, et al: Glycosylated minor components of human adult hemoglobin. *J Biol Chem* 1978;253:2327–2332.

McFarland KF, Catalano EW, Day JF, et al: Nonenzymatic glucosylation of serum proteins in diabetes mellitus. *Diabetes* 1979;28:1011–1014.

McFarland KF, Murtiashaw M, Baynes JW: Clinical values of glycosylated serum protein and glycosylated hemoglobin levels in the diagnosis of gestational diabetes mellitus. *Obstet Gynecol* 1984;64:516–518.

McKeown-Longo PJ, Mosher DF: Binding of plasma fibronectin to cell layers of human skin fibroblasts. *J Cell Biol* 1983;97:466–472.

McVerry BA, Fisher C, Hopp A, et al: Production of pseudo-diabetic renal glomerular changes in mice after repeated injections of glucosylated proteins. *Lancet* 1980;1:738–740.

McVerry BA, Thorpe S, Joe F, et al: Nonenzymatic glucosylation of fibrinogen. *Haemostasis* 1981;10:261–270.

Mecklenberg RS, Benson EA, Benson JW Jr, et al: Long-term metabolic control with insulin pump therapy: Report of experience with 127 patients. *N Engl J Med* 1985;313:465–468.

Mehl TD, Wenzel SE, Russell B, et al: Comparison of two indices of glycemic control in diabetic subjects: Glycosylated serum protein and hemoglobin. *Diabetes Care* 1983;6:34–39.

Menard L, Dempsey ME, Blankstein LA, et al: Quantitative determination of glycosylated hemoglobin A_1 by agar gel electrophoresis. *Clin Chem* 1980;26:1598–1602.

Michael AF, Brown DM: Increased concentration of albumin in kidney basement membranes in diabetes mellitus. *Diabetes* 1981;308:843–846.

Miller JM Jr, Crenshaw C Jr, Welt SI: Hemoglobin A_{1c} in normal and diabetic pregnancy. 02JAMA 1979;242:2785–2787.

Miller E, Hare JW, Cloherty JP, et al: Elevated maternal hemoglobin A_{1c} in early pregnancy and major congenital anomalies in infants of diabetic mothers. *N Engl J Med* 1981;304:1331–1334.

Miller K, Michael AF: Immunopathology of renal extracellular membranes in diabetes: Specificity of tubular basement membrane immunofluorescence. *Diabetes* 1976;25:701–708.

Mohammad A, Olcott HS, Fraenkel-Conrat H: The "browning" reaction of proteins with glucose. *Arch Biochem* 1979;24:157–163.

Monnier VM, Cerami A: Nonenzymatic browning in vivo: Possible process for aging of long-lived proteins. *Science* 1981;211:491–493.

Monnier VM, Cerami A: The search for non-enzymatic browning products in the human lens. *Invest Ophthalmol Vis Sci* 1981;20:169.

Monnier VM, Cerami A: Nonenzymatic glycosylation and browning of proteins in diabetes. *Clin Endocrinol Metab* 1982;11:431–452.

Monnier VM, Cerami A: Detection of nonenzymatic browning products in the human lens. *Biochim Biophys Acta* 1983;760:97–103.

Monnier VM, Kohn RR, Cerami A: Accelerated age-related browning of human collagen in diabetes mellitus. *Proc Natl Acad Sci USA* 1984;81:583–587.

Monnier VM, Stevens VJ, Cerami A: Nonenzymatic glycosylation, sulfhydryl oxidation, and aggregation of lens proteins in experimental sugar cataracts. *J Exp Med* 1979;150:1098–1107.

Moore WV, Knapps J, Kauffman RL, et al: Plasma lipid levels in insulin-dependent diabetes mellitus. *Diabetes Care* 1979;2:31–34.

Mortensen HB: Quantitative determination of hemoglobin A_{1c} by thin-layer isoelectric focusing. *J Chromatogr* 1980;182:325–333.

Mosher DF: Physiology of fibronectin. *Annu Rev Med* 1984;35:561–565.

Mosher DF, Sakesla O, Keski-Oja J, et al: Distribution of a major surface-associated glycoprotein, fibronectin, in cultures of adherent cells. *J Supramol Struct* 1977;6:551–557.

Nahum HD, Lonchampt MM, Duhault J, et al: Nonenzymatic plasma glycosylation: Lack of saturation at high glucose concentration. *IRCS Med Sci* 1982;10:436.

Nathan DM: Labile glycosylated hemoglobin contributes to hemoglobin A_1 as measured by liquid chromatography or electrophoresis. *Clin Chem* 1981;27:1261–1263.

Nathan DM, Avezzano ES, Palmer JL: A rapid chemical means for removing labile glycohemoglobin. *Diabetes* 1981;30:700–701.

Nathke HE, Siess EA, Wieland OH: Glucosylated plasma protein injection does not produce glomerular basement membrane thickening. *Horm Metab Res* 1984;16:557–558.

Neglia CI, Cohen HJ, Garber AR, et al: ^{13}C NMR investigation of non-enzymatic glucosylation of protein: Model studies using RNase A. *J Biol Chem* 1983;258:14279–14283.

Ney KA, Colley KJ, Pizzo SV: The standardization of the thiobarbituric acid assay for nonenzymatic glucosylation of human serum albumin. *Anal Biochem* 1981;118:294–300.

Ney KA, Pasqua JJ, Colley KJ, et al: In vitro preparation of nonenzymatically glucosylated human transferrin, α_2-macroglobulin, and fibrinogen with preservation of function. *Diabetes* 1985;34:462–470.

Nicol DJ, Davis RE, Curnow DH: A simplified colorimetric method for the measurement of glycosylated hemoglobin. *Pathology* 1983;15:443–447.

Nicol DJ, Davis RE, McCann VJ: Hemoglobinopathy in patients with diabetes mellitus; A complicating factor in the measurement of glycosylated hemoglobin, letter to the editor. *Diabetes Care* 1983;6:524.

Nikkila EA, Hormila P: Serum lipids and lipoproteins in insulin-treated diabetes. *Diabetes* 1976;27:1078–1086.

Oberly TD, Mosher DF, Mills MD: Localization of fibronectin within the renal glomerulus and its production by cultured glomerular cells. *Am J Pathol* 1979;96:651–662.

Oberly TD, Murphy-Ullrich JE, Albrecht RM, et al: The effect of the dimeric and multimeric forms of fibronectin on the adhesion and growth of primary glomerular cells. *Exp Cell Res* 1983;145:265-276.

Oimomi M, Hatanaka H, Ishikawa K, et al: Increased fructose-lysine of nail protein in diabetic patients. *Klin Wochenschr* 1984;62:477-478.

Oimomi M, Yoshimura Y, Kubota S, et al: Hemoglobin A_1 properties of diabetic and uremic patients. *Diabetes Care* 1981;4:484-486.

Ogata ES, Freinkel N, Metzger BE, et al: Perinatal islet function in gestational diabetes: Assessment by cord plasma: C-peptide and amniotic fluid insulin. *Diabetes Care* 1980;3:425-429.

O'Shaughnessy R, Russ J, Zuspan FP: Glycosylated hemoglobins and diabetes mellitus in pregnancy. *Am J Obstet Gynecol* 1979;135:783-790.

Paisey R, Pennoch C, Owens C, et al: Rapid glycosylation of hemoglobin. *Diabetologia* 1981;20:80.

Pande A, Garner WH, Spector A: Glucosylation of human lens protein and cataractogenesis. *Biochem Biophys Res Commun* 1979;89:1260-1266.

Parathasarathy N, Spiro RG: Effect of diabetes on the glycosaminoglycan component of the human glomerular basement membrane. *Diabetes* 1982;31: 738-741.

Parker KM, England JD, DaCosta J, et al: Improved colorimetric assay for glycosylated hemoglobin. *Clin Chem* 1981;27:669-672.

Paulsen EP, Khoury M: Hemoglobin A_{1c} levels in insulin-dependent and independent diabetes mellitus. *Diabetes* 1976;25:890-896.

Pecoraro RE, Chen MS, Porte D: Glycosylated hemoglobin and fasting plasma glucose in the assessment of outpatient glycemic control in NIDDM. *Diabetes Care* 1982;5:592-599.

Pecoraro RE, Graf RJ, Halter JB, et al: Comparison of a colorimetric assay for glycosylated hemoglobin with ion exchange chromatography. *Diabetes* 1979;28:1120-1125.

Pedersen J, Bojsen-Moller B, Poulsen H: Blood sugar in newborn infants of diabetic mothers. *Acta Endocrinol* 1954;15:33-36.

Pedersen JF, Molsted-Pedersen L, Mortensen HB: Fetal growth delay and maternal hemoglobin A_{1c} in early diabetic pregnancy. *Obstet Gynecol* 1984;64:351-352.

Perejda AJ, Uitto J: Nonenzymatic glycosylation of collagen and other proteins. Relationship to development of diabetic complications. *Coll Rel Res* 1982;2:81-88.

Perejda AJ Zargoza EJ, Eriksen E, et al: Nonenzymatic glucosylation of lysyl and hydroxylysyl residues in Type I and Type II collagens. *Coll Rel Res* 1984;4:427-439.

Peterson CM, Jones RL, Esterly JA, et al: Changes in basement membrane thickening and pulse volumes concomitant with improved glucose control and exercise in patients with insulin-dependent diabetes mellitus. *Diabetes Care* 1980;3:586-589.

Peterson CM, Koenig RJ, Jones RL, et al: Correlation of serum triglyceride levels and Hb A_{1c} concentrations in diabetes mellitus. *Diabetes* 1977;26:507-509.

Pinnell SR, Martin GR: The cross-linking of collagen and elastin: Enzymatic conversion of lysine and peptide linkage to α-aminoadipic-δ-semialdehyde (allysine) by an extract from bone. *Proc Natl Acad Sci USA* 1968;61:708–716.

Pirie A: Color and solubility of the proteins of human cataracts. *Invest Ophthalmol* 1968;7:634–650.

Pongor S, Ulrich PC, Bencsath A, et al: Aging of proteins: Isolation and identification of a fluorescent chromophore from the reaction of polypeptides with glucose. *Proc Natl Acad Sci USA* 1984;81:2684–2688.

Puukka R, Hekali R, Akerblom HK, et al: Haemoglobin Hijiyama: Haemoglobin variant found in connection with glycosylated haemoglobin estimation in a Finnish diabetic boy. *Clin Chim Acta* 1982;121:51–57.

Rabhar S: An abnormal hemoglobin in red cells of diabetics. *Clin Chim Acta* 1968;22:296–298.

Rabhar S, Blumenfeld O, Ranney HM: Studies of the unusual hemoglobin in patients with diabetes mellitus. *Biochem Biophys Res Commun* 1969;36:838–843.

Rao GN, Lardis MP, Cotlier E: Acetylation of lens crystallins: A possible mechanism by which aspirin could prevent cataract formation. *Biochem Biophys Res Commun* 1985;128:1125–1132.

Raskin P, Pietri AO, Unger R, et al: The effect of diabetic control of the width of skeletal-muscle capillary basement membrane in patients with Type I diabetes mellitus. *N Engl J Med* 1983;309:1546–1550.

Rendell M, Kao G, Micherikunnel P, et al: Use of aminophenylboronic acid affinity chromatography to measure glycosylated albumin levels. *J Lab Clin Med* 1985;105:63–69.

Rendell M, Stephen PM, Paulsen R, et al: An interspecies comparison of normal levels of glycosylated hemoglobin and glycosylated albumin. *Comp Biochem Physiol* 1985;81B:819–822.

Reynolds TM: Chemistry of nonenzymatic browning. *Adv Food Res* 1963;12:1–52.

Reynolds TM: Chemistry of nonenzymatic browning. *Adv Food Res* 1965;14:167–283.

Robins SP, Bailey AJ: Age-related changes in collagen: The identification of reducible lysine-carbohydrate condensation products. *Biochem Biophys Res Commun* 1972;48:76–84.

Rogozinski S, Blumenfeld OO, Seifter S: The nonenzymatic glycosylation of collagen. *Arch Biochem Biophys* 1983;221:428–437.

Rosenberg H, Modrak JB, Hassing JM, et al: Glycosylated collagen. *Biochem Biophys Res Commun* 1979;91:498–501.

Rucklidge GJ, Bates GP, Robins SP: Preparation and analysis of the products of non-enzymatic protein glycosylation and their relationship to cross-linking of proteins. *Biochim Biophys Acta* 1983;747:165–170.

Ruoslahti E, Engvall E, Hagman EG: Fibronectin: Current concepts of its structure and function. *Collagen Res* 1981;1:95–128.

Saibene V, Brembilla L, Bertoletti A, et al: Chromatographic and colorimetric detection of glycosylated hemoglobins: A comparative analysis of two different methods. *Clin Chim Acta* 1979;93:199–205.

Sakurai T, Takahasi H, Tsuchiya S: New fluorescence of nonenzymatically glucosylated human serum albumin. *FEBS Lett* 1984;176:27–31.

Saltmarch M, Labuzza TP: Nonenzymatic browning via the Maillard reaction in foods. *Diabetes* 1982;31(suppl 3):29–36.

Samaja M, Melotti D, Carenini A, et al: Glycosylated haemoglobins and the oxygen affinity of whole blood. *Diabetologia* 1982;23:399–402.

Santiago JV, Davis JE, Fisher F: Hemoglobin A_{1c} levels in a diabetes detection program. *J Clin Endocrinol Metab* 1978;47:578–580.

Saraswathi S, Colman RW: Role of galactose in bovine factor V. *J Biol Chem* 1975;250:8111–8118.

Sasaki J, Arora V, Cottam GL: Nonenzymatic galactosylation of human LDL decreases its metabolism by human skin fibroblasts. *Biochem Biophys Res Commun* 1982;108:791–796.

Sasaki J, Cottam GL: Glycosylation of LDL decreases its ability to interact with high-affinity receptors of human fibroblasts in vitro and decreases its clearance from rabbit plasma in vivo. *Biochem Biophys Acta* 1982;713:199–207.

Sasaki J, Cottam GL: Glycosylation of human LDL and its metabolism in human skin fibroblasts. *Biochem Biophys Res Commun* 1982;104:977–983.

Sasaki J, Okamura T, Cottam GL: Measurement of receptor-independent metabolism of low-density lipoprotein: An application of glycosylated low-density lipoprotein. *Eur J Biochem* 1983;131:535–538.

Schleicher E, Deufel T, Wieland OH: Non-enzymatic glycosylation of human serum lipoproteins: Elevated ε-lysine glycosylated low density lipoprotein in diabetic patients. *FEBS Lett* 1981;129:1–4.

Schleicher ED, Gerbitz KD, Dolhofer R, et al: Clinical utility of non-enzymatically glycosylated blood proteins as an index of glucose control. *Diabetes Care* 1984;7:548–558.

Schleicher E, Olgemöller B, Schön J, et al: Limited nonenzymatic glucosylation of low-density lipoprotein does not alter its catabolism in tissue culture. *Biochim Biophys Acta* 1985;846:226–233.

Schleicher E, Wieland OH: Specific quantitation by HPLC of protein (lysine) bound glucose in human serum albumin and other glycosylated proteins. *J Clin Chem Clin Biochem* 1981;19:81–87.

Schleicher E, Wieland OH: Changes of human glomerular basement membrane in diabetes mellitus. *J Clin Chem Clin Biochem* 1984;22:223–237.

Schnek AG, Schroeder WA: The relation between the minor components of whole normal adult hemoglobin as isolated by chromatography and starch block electrophoresis. *J Am Chem Soc* 1961;83:1472–1478.

Schnider SL, Kohn RR: Glucosylation of human collagen in aging and diabetes mellitus. *J Clin Invest* 1980;66:1179–1181.

Schnider SL, Kohn RR: Effects of age and diabetes mellitus on the solubility and nonenzymatic glucosylation of human skin collagen. *J Clin Invest* 1981;67:1630–1635.

Schultz TA, Lewis SB, Davis JL, et al: Effect of sulfonylurea therapy and plasma glucose levels on hemoglobin A_{1c} in Type II diabetes mellitus. *Am J Med* 1981;70:373–378.

Schwartz B, Gray GR: Proteins containing reductively aminated disaccharides. *Arch Biochem Biophys* 1977;181:542-549.

Schwartz HC, King KC, Schwartz AL, et al: Effects of pregnancy on hemoglobin A_{1c} in normal, gestational diabetic, and diabetic women. *Diabetes* 1976;25:1118-1122.

Schwimmer S, Olcott HS: Reaction between glycine and the hexose phosphates. *J Am Chem Soc* 1953;75:4855-4856.

Seibold JR, Uitto J, Dorwart BB, et al: Collagen synthesis and collagenase activity in dermal fibroblasts from patients with diabetes and digital sclerosis. *J Lab Clin Med* 1985;105:664-667.

Service FJ, Fairbanks VF, Rizza RA: Effect on hemoglobin $dses_1$ of rapid normalization of glycemia with an artifical endocrine pancreas. *Mayo Clin Proc* 1981;56:377-380.

Shaklai N, Garlick RL, Bunn HF: Nonenzymatic glycosylation of human serum albumin alters its conformation and function *J Biol Chem* 1984; 259:3812-3817.

Shapiro R, McManus MJ, Zalut C, et al: Sites of nonenzymatic glycosylation of human hemoglobin A. *J Biol Chem* 1980;255:3120-3127.

Siegel RC, Martin GR: Collagen cross-linking: Enzymatic synthesis of lysine-derived aldehydes and the production of cross-linked components. *J Biol Chem* 1970;245:1653-58.

Simon D, Coignet MC, Thibult N, et al: Comparison of glycosylated hemoglobin and fasting plasma glucose with two-hour post-load plasma glucose in the detection of diabetes mellitus. *Am J Epidemiol* 1985;122:589-593.

Simon M, Cuan J: Hemoglobin A_{1c} by isoelectric focusing. *Clin Chem* 1982;28:9-12.

Smith RJ, Koenig RJ, Binnerts A, et al: Regulation of hemoglobin A_{1c} formation in human erythrocytes in vitro: Effects of physiologic factors other than glucose. *J Clin Invest* 1982;69:1164-1168.

Soler NG, Frank S: Value of glycosylated hemoglobin measurements after myocardial infarction. *JAMA* 1981;246:1690-1693.

Solway J, McDonald M, Bunn HF, et al: Biosynthesis of glycosylated hemoglobin in the monkey. *J Lab Clin Med* 1979;93:962-973.

Sosenko JM, Breslow JL, Muttinew OS, et al: Hyperglycemia and plasma lipid levels: A prospective study of young insulin-dependent diabetic patients. *N Engl J Med* 1980;302:650-654.

Sosenko JM, Flückiger R, Platt OS, et al: Glycosylation of variant hemoglobins in normal and diabetic subjects. *Diabetes Care* 1980;3:590-593.

Sosenko JM, Kitzmiller JL, Flückiger R, et al: Umbilical cord glycosylated hemoglobin in infants of diabetic mothers: Relationship to neonatal hypoglycemia, macrosomia, and cord serum C-peptide. *Diabetes Care* 1982;5:566-570.

Sosenko IR, Kitzmiller JL, Loo SW, et al: The infant of the diabetic mother: Correlation of increased cord C-peptide levels with macrosomia and hypoglycemia. *N Engl J Med* 1979;301:859-862.

Sosenko JM, Miettinen OS, Williamson JR, et al: Muscle capillary basement membrane thickness and long-term glycemia in Type I diabetes millitus. *N Engl J Med* 1984;311:694–698.

Sowers JR, Tuck ML, Sowers DK: Plasma antithrombin III and thrombin generation time: Correlation with hemoglobin A and fasting serum glucose in young diabetic women. *Diabetes Care* 1980;3:655–658.

Spicer KM, Allen RC, Buse MG: Simplified assay of hemoglobin A_{1c} in diabetic patients by use of isoelectric focusing and quantitative micro-densitometry. *Diabetes* 1978;27:384–388.

Standl E, Kolb HJ: 2,3-Diphosphoglycerate fluctuations in erythrocytes reflecting pronounced blood glucose variance. In vivo and in vitro studies in normal, diabetic and hypoglycemic subjects. *Diabetologia* 1973;9:461–466.

Starkman HS, Wacks M, Soeldner JS, et al: Effect of acute blood loss on glycosylated hemoglobin determinations in normal subjects. *Diabetes Care* 1983;6:291–294.

Steinbrecher VP, Witztum JL: Glucosylation of low-density lipoproteins to an extent comparable to that seen in diabetes slows their catabolism. *Diabetes* 1984;33:130–134.

Stenman S, Vaheri A: Distribution of a major connective tissue protein, fibronectin, in normal human tissues. *J Exp Med* 1978;147:1054–1064.

Stevens VJ, Monnier VM, Cerami A: Hemoglobin glycosylation as a model for modification of other proteins. *Texas Rep Biol Med* 1980-1981;40:387–396.

Steven VJ, Fantl WJ, Newman CB, et al: Acetaldehyde adducts with hemoglobin. *J Clin Invest* 1981;67:361–369.

Stevens VJ, Rouzer CA, Monnier VM, et al: Diabetic cataract formation: Potential role of glycosylation of lens crystallins. *Proc Natl Acad Sci USA* 1978;75:2918–922.

Stevens VJ, Vlassara H, Abati A, et al: Nonenzymatic glycosylation of hemoglobin. *J Biol Chem* 1977;252:2998–3002.

Stickland MH, Perkins CM, Wales JK: The measurement of haemoglobin A_{1c} by isoelectric focussing in diabetic patients. *Diabetologia* 1982;22:315–317.

Subramanian CV, Radhakrishnamurthy B, Berenson GS: Photometric determination of glycosylation of hemoglobin in diabetes mellitus. *Clin Chem* 1980;26:1683–1687.

Summerfield JA, Vergalla J, Jones E: Modulation of a glycoprotein recognition system on rat hepatic endothelial cells by glucose and diabetes mellitus. *J Clin Invest* 1982;69:1337–1347.

Susa JB, McCormick KL, Widness JA, et al: Chronic hyperinsulinemia in the fetal rhesus monkey: Effects on fetal growth and composition. *Diabetes* 1979;28:1058–1063.

Susa JB, Neave C, Sehgal P, et al: Chronic hyperinsulinemia in the fetal rhesus monkey: Effect of physiologic hyperinsulinemia on fetal growth and composition. *Diabetes* 1984;33:656–660.

Susa JB, Schwartz R: Effects of hyperinsulinemia in the primate fetus. *Diabetes* 1985;34(suppl 2):36–41.

Susa JB, Widness JA, Hintz R, et al: Somatomedins and insulin in diabetic pregnancies: Effects on fetal macrosomia in the human and rhesus monkey. *J Clin Endocrinol Metab* 1984;58:1099–1105.

Svendsen PA, Christiansen JS, Andersen RA, et al: Fast glycosylation of haemoglobin, letter to the editor. *Lancet* 1979;1:1142–1143.

Svendsen PA, Christiansen JS, Soegaard U, et al: Rapid changes in chromatographically determined haemoglobin A_{1c} induced by short term changes in glucose concentration. *Diabetologia* 1980;19:130–136.

Svendsen PA, Christiansen JS, Welinder B, et al: Fast glycosylation of haemoglobin, letter to the editor. *Lancet* 1979;1:603.

Tanzer ML: Isolation of lysinonorleucine from collagen. *Biochem Biophys Res Commun* 1970;39:183–189.

Tanzer ML: Cross-linking of collagen: Endogenous aldehydes react in several ways to form a variety of unique covalent cross-links. *Science* 1973;180:561–566.

Tanzer ML, Fairweather R, Gallop PM: Collagen cross-links: Isolation of reducible Nε-hexosyl hydroxylysine from borohydride reduced calf skin insoluble collagen. *Arch Biochem Biophys* 1972;48:76–84.

Tanzer ML, Kefalides NA: Collagen cross-links: Occurrence in basement membrane collagens. *Biochem Biophys Res Commun* 1973;51:775–780.

Tarsio JF, Wigness B, Rhode TD, et al: Nonenzymatic glycation of fibronectin and alterations in the molecular association of cell matrix and basement membrane components in diabetes mellitus. *Diabetes* 1985;34:477–484.

Thornton WE, Schellekens APM, Sanders GTB: Assay of glycosylated hemoglobin using agar electrophoresis. *Ann Clin Biochem* 1981;18:182–184.

Travis J, Bowen J, Tewksbury D, et al: Isolation of albumin from whole human plasma and fractionation of albumin-depleted plasma. *Biochem J* 1976;157:301–306.

Travis J, Pannell R: Selective removal of albumin from plasma by affinity chromatography. *Clin Chim Acta* 1973;49:49–52.

Trivelli LA, Ranney HM, Lai HT: Hemoglobin components in patients with diabetes mellitus. *N Engl J Med* 1971;284:353–357.

Trüeb B, Flückiger R, Winterhalter KH: Nonenzymatic glycosylation of basement membrane collagen in diabetes mellitus. *Coll Rel Res* 1984;4:239–251.

Trüeb B, Holenstein CG, Fischer RW, et al: Nonenzymatic glycosylation of proteins: A warning. *J Biol Chem* 1980;255:6717–6720.

Trüeb B, Hughes GJ, Winterhalter KH: Synthesis and quantitation of glucitollysine, a glycosylated amino acid elevated in proteins from diabetics. *Anal Biochem* 1982;119:330–334.

Tsuchiya S, Sakurai T, Sekiguchi S-L: Nonenzymatic glucosylation of human serum albumin and its influence on binding capacity of sulfonylureas. *Biochem Pharmacol* 1984;33:2967–2971.

Turner RC, et al (a multicenter study): U.K. Prospective Diabetes Study: II. Reduction in HbA_{1c} with basal insulin supplement, sulfonyurea, or biguanide therapy in maturity-onset diabetes. *Diabetes* 1985;34:793–798.

Uitto J, Perejda AJ, Grant GA, et al: Glycosylation of human glomerular basement membrane collagen: Increased content of hexose in ketoamine linkage and unaltered hydroxylysine-glycosides in patients with diabetes. *Connect Tiss Res* 1982;10:287-296.

Urbanowski JC, Cohenford MA, Dain JA: Nonenzymatic galactosylation of human serum albumin. *J Biol Chem* 1982;257:111-115.

Urdanivia E, Cohen MP: Identification of amino acid sites of [^{14}C]-glucosylation of basement membranes. *Clin Res* 1981;29:734A.

Vlassara H, Brownlee M, Cerami A: Nonenzymatic glycosylation of peripheral nerve protein in diabetes mellitus. *Proc Natl Acad Sci USA* 1981;78:5190-5192.

Vlassara H, Brownlee M, Cerami A: Excessive nonenzymatic glycosylation of peripheral and central nervous system myelin components in diabetic rats. *Diabetes* 1983;32:670-674.

Vlassara H, Brownlee M, Cerami A: Accumulation of diabetic rat peripheral nerve myelin by macrophages increases with the presence of advanced glycosylation end products. *J Exp Med* 1984;160:197-207.

Vlassara H, Brownlee M, Cerami A: High-affinity-receptor-mediated uptake and degradation of glucose-modified proteins: A potential mechanism for the removal of senescent macromolecules. *Proc Natl Acad Sci USA* 1985;82:5588-5592.

Vlassara H, Brownlee M, Cerami A: Recognition and uptake of human diabetic peripheral nerve myelin by macrophages. *Diabetes* 1985;34:553-557.

Vogt BW, Schleicher ED, Wieland OH: ε-Amino-lysine bound glucose in human tissues obtained at autopsy: Increase in diabetes mellitus. *Diabetes* 1982;31:1123-1127.

Walinder O, Wibell L, Tuvemo T: Relation between hemoglobin A and determinations of glucose in diabetics treated with and without insulin. *Diabetes Metab* 1980;6:251-255.

Walton DJ, Ison ER, Szarek WA: Synthesis of *N*-(1-deoxy-hexitol-1-yl) amino acids, reference compounds for the nonenzymatic glycosylation of proteins. *Carbohydr Res* 1984;128:37-49.

Welch SG, Boucher BJ: A rapid micro-scale method for the measurement of hemoglobin A$_1$ (a+b+c). *Diabetologia* 1978;14:209-211.

White NJ, Waltman SR, Krupin T, et al: Reversal of early ocular abnormalities in juvenile diabetics (IDD) after normalization of hemoglobin A$_{1c}$. *Clin Res* 1981;426A.

Widness JA, Rogler-Brown TL, McCormick KL, et al: Rapid fluctuations in glycohemoglobin (hemoglobin A$_{1c}$) related to acute changes in glucose. *J Lab Clin Med* 1980;95:386-394.

Widness JA, Schwartz HC, Kahn CB, et al: Glycohemoglobin in diabetic pregnancy: A sequential study. *Am J Obstet Gynecol* 1980;136:1024-1029.

Widness JA, Schwartz HC, Thompson D, et al: Glycohemoglobin (Hb A$_{1c}$): A predictor of birth weights in infants of diabetic women. *J Pediatr* 1978;92:8-12.

Willey DG, Rosenthal MA, Caldwell S: Glycosylated haemoglobin and plasma glycoprotein assays by affinity chromatography. *Diabetologia* 1984;27:56-58.

Williams SK, Devenney JJ, Bitensky MW: Micropinocytic ingestion of glycosylated albumin by isolated microvessels; Possible role in pathogenesis of diabetic microangiopathy. *Proc Natl Acad Sci USA* 1981;78:2393–2397.

Williams JH, Hillson RM, Bron A, et al: Retinopathy is associated with higher glycemia in maturity-onset diabetes. *Diabetologia* 1984;27:198–202.

Williams SK, Howarth NL, Devenney JJ, et al: Structural and functional consequences of increased tubulin glycosylation in diabetes mellitus. *Proc Natl Acad Sci USA* 1982;79:6546–6550.

Williams SK, Siegal RK: Preferential transport of nonenzymatically glucosylated ferritin across the kidney glomerulus. *Kidney Int* 1985;28:146–152.

Williams SK, Solenski NJ: Enhanced vesicular ingestion of nonenzymatically glucosylated proteins by capillary endothelium. *Microvasc Res* 1984;28:311–321.

Winterhalter KH, Glatthaar B: Chromatographic separation of Hb A_{1c}. *Semin Haematol* 1971;4:84–96.

Wiseman MJ, Saunders AJ, Keen H, et al: Effect of blood glucose control on increased glomerular filtration rate and kidney size in insulin-dependent diabetes. *N Engl J Med* 1985;312:617–621.

Witztum JL, Mahoney EM, Branks MJ, et al: Nonenzymatic glucosylation of low-density lipoproteins alters its biologic activity. *Diabetes* 1982;31:283–291.

Wu V-Y, Cohen MP: Reducible cross-links in human glomerular basement membrane. *Biochem Biophys Res Commun* 1982;104:911–915.

Yamada KM: Cell surface interactions with extracellular materials. *Annu Rev Biochem* 1983;52:761–799.

Yamada KM, Kennedy DW, Kimata K, et al: Characterization of fibronectin interactions with glycosaminoglycans and identification of active proteolytic fragments. *J Biol Chem* 1980;255:6055–6063.

Ylinen K, Aula P, Stenman UH, et al: Risk of minor and major fetal malformations in diabetics with high haemoglobin A_{1c} values in early pregnancy. *Br Med J* 1984;289:345–346.

Yosha SF, Elden HR, Rabinovitch A, et al: Experimental diabetes mellitus and age-stimulated changes in intact rat dermal collagen. *Diabetes* 1983;32:739–742.

Yue DK, McLennan S, Church DB, et al: The measurement of glycosylated hemoglobin in man and animals by aminophenylboronic acid affinity chromatography. *Diabetes* 1982;31:701–707.

Yue DK, McLennan S, Delbridge L, et al: The thermal stability of collagen in diabetic rats: Correlation with severity of diabetes and nonenzymatic glycosylation. *Diabetologia* 1983;24:282–285.

Yue DK, McLennan S, Handelsman DJ, et al: The effect of salicylates on nonenzymatic glycosylation and thermal stability of collagen in diabetic rats. *Diabetes* 1984;33:745–751.

Yue DK, Morris, K, McLennan S, et al: Glycosylation of plasma protein and its relation to glycosylated hemoglobin in diabetes. *Diabetes* 1980;29:296–300.

Index